Principles of

Medical Electronics *and* Biomedical Instrumentation

Principles of Medical Electronics *and* Biomedical Instrumentation

C Raja Rao
S K Guha

Universities Press

Universities Press (India) Private Limited

Registered Office
3-6-747/1/A & 3-6-754/1, Himayatnagar,
Hyderabad 500 029 (A.P.), India
e-mail: info@universitiespress.com

Distributed by
Orient Blackswan Private Limited

Registered Office
3-6-752, Himayatnagar, Hyderabad 500 029 (A.P), India

Other Offices
Bangalore / Bhopal / Bhubaneswar / Chennai
Ernakulam / Guwahati / Hyderabad / Jaipur / Kolkata
Lucknow / Mumbai / New Delhi / Patna

First published 2001
Reprinted 2005, 2007, 2009, 2013

ISBN: 978 81 7371 257 9

Typeset by
Venture Graphics
Chennai 600 031

Printed in India at
Bluebells Secure Prints
Hyderabad 500 072

Published by
Universities Press (India) Private Limited
3-6-747/1/A & 3-6-754/1, Himayatnagar,
Hyderabad 500 029 (A.P.), India

Contents

Foreword

The integration of concepts and techniques drawn from various disciplines is the key to advancement in modern science and technology. Particularly the living system with all its complexities, presents unique challenges in understanding the mechanisms of action and applying the knowledge for the preservation of its smooth functioning. Therefore, all human endeavours towards good health must begin with an effort to bring together, on one platform, all that which is needed from relevant areas. It is a difficult task considering the volume of information existing today. Professors C Raja Rao and Sujoy K Guha with their unique backgrounds in engineering, basic medical science and clinical medicine have addressed the issue in a truly special manner. Using a multifaceted approach, the latest frontiers of biomedical knowledge have been interdigitated with current technology, especially electronics, to give the essentials as well as a holistic picture of technology in medical practice. Therefore, the book of *Principles of Medical Electronics and Biomedical Instrumentation* will be of immense value both as a textbook and as a reference for students and professionals in engineering and biomedicine. It is a great pleasure to introduce to the engineering and medical communities an outcome of the excellent and dedicated efforts of two leading academicians, supported by a grant from the Indian Institute of Technology Delhi.

Professor V S Raju
Director
Indian Institute of Technology (IIT)
New Delhi

Preface

We are now living in the Age of Electronics; electronics has revolutionized global communications and the field of entertainment through satellite television. Now the "Diagnosticelectron" is poised for extensive use in hospitals and nursing homes for better health care. Modern medicine is so much centered around electronics, that the doctor of the future must surely also be an engineer. In fact, it would not be an exaggeration to say that the future doctor must be well grounded both in engineering and mathematics.

This book is a description of the medical and technical disciplines that make up this new world of "Medical Electronics". It should prove useful as a reference for medical and paramedical personnel who have some knowledge of instruments and who need to know more. For the electronics engineer and technician, the book will serve as a bridge to the medical world they will be serving. For the administrator, it will be an explanation of his new department and what it should accomplish.

Medical electronics is a part of a larger area of Biomedical Engineering and, lately, remarkable progress has taken place in medical engineering in general and medical electronics, in particular, as an aid in the alleviation of human suffering. Perhaps the three major impacts on biomedical engineering in recent years are the tremendous expansion of non-invasive techniques, the sophistication built up in special care units and, along with other fields, the greater use of computers and the advent of microprocessors. A wide range of medical electronic equipment for purposes of medical diagnosis, therapy, research and bioanalytical investigations are already being manufactured and marketed.

After teaching the subject matter to undergraduate and postgraduate students for more than two decades and having guided a few students for Ph D degrees, the authors ventured to write this book of 18 chapters in order to meet a multiplicity of needs they had encountered.

The book is written primarily to cater to the syllabi of Medical Electronics/ Biomedical Engineering, taught as a core elective subject to the B Tech and M Tech students in various branches of Engineering, Applied Physics and other allied fields, in addition to serving the needs of A M I E (Section B) students. Emphasis has, therefore, been laid more on the fundamentals than on detailed technical descriptions. Being an interdisciplinary field, an effort has been made to provide some biomedical knowledge as well as the technology of instrumentation.

Chapter 1 gives an introduction to the subject, thus providing the reader a perspective of the field and a feeling for the subject matter. Chapter 2 deals with various aspects of the human cell including the structural and electrical properties, resting and action potentials, Nernst equation, etc. The physiological aspects of the cardiovascular system, central nervous system, muscle action and sensory and motor nerves, finally giving the basis to the biopotentials — ECG, EEG, and EMG which are considered essential, before dealing with various concepts of biomedical instrumentation are briefly described in Chapter 3.

Chapters 4, 5, 6 and 7 are devoted to the four basic stages of measurements and instrumentation, i.e., bioelectrodes, transducers for biomedical applications, bioamplifiers and principles of recorders for recording bioelectric events. Having given the explanation that the human body is a source of a variety of biosignals, highly significant for diagnosis and therapy, enunciating the characteristics of the devices for their general measurement. Chapters 8 to 10 highlight the specific measurement techniques and apparatus.

In Chapter 8, dealing with electrocardiography, measurements on various cardiac planes are covered, concluding with the description and working of the electrocardiograph. Similarly, Chapter 9 deals with the EEG recording modes and techniques, applications of EEG, culminating in the description and functioning of the electroencephalograph. The EMG recording system and determination of nerve conduction velocities are described in Chapter 10, besides exposing the reader specifically to electrical activity of the eye in terms of ERG and EOG and that of stomach EGG, very briefly.

Chapter 11 gives concisely the mechanism, basics of measurements on lung volume, air flow rate and tidal volume. Special techniques for measurement of non-electrical biological parameters like electrical impedance plethysmography, audiometry, diagnostic and therapeutical aspects of X-rays, diagnostic ultrasound and blood flow measurement are presented in Chapter 12.

The 13th chapter discusses a group of medical electronic equipment, which are essentially therapeutic and prosthetic in nature. These devices affect the body or provide functional support and, at times, may even be crucial for survival. They include defibrillators, pacemakers, various types of diathermy apparatus, blood pumps, respirators, and electronic systems for paralysed muscles. Chapter 14, on biotelemetry, briefly explains the basic techniques for monitoring physiological data over a distance wirelessly, and also the usage of telephone lines for short distances.

In modern hospitals, continuous and automatic patient monitoring and prompt care and precautions taken thereon to meet an emergency situation are briefly outlined in Chapter 15. With the increasing use of monitoring and therapeutical electromedical equipment, the patient has been included as a part of an electrical circuit and thus exposed to a variety of electrical hazards. Chapter 16 gives a concise idea of electrical shock, its physiological effects, preventive measures to reduce shock hazards and steps taken to maintain a safe patient electrical environment. Medicine is one field where the whole range of computer power can be, and indeed in many cases is being, used to improve the standards of diagnosis and treatment of diseases without decreasing the need for the skills of highly-trained clinical and nursing staff. The range of computer applications in medicine is so great that only a few key areas are briefly analyzed in Chapter 17. Medical electronics is a large factor in the total operation of the hospital's basic role of treating patients. Thus, the need to establish a "Medical Electronics Department" in a modern hospital, its responsibilities and its professionalism and other commitments are voiced in Chapter 18.

The authors wish to thank Mr A S Babu immensely for his assistance in typing the manuscript and computer processing.

Dr Sujoy K Guha
Dr C Raja Rao
September 2000

Chapter 1

Introduction

1.1 Electronics and Medicine

Electronics in the hospital is a "hot" subject today. For the manufacturing community, it is a new market. For the researcher, it is an opportunity to try out what were once only ideas. For the physician, it is an additional tool. For the engineer, it is a new area for applying technology. For the administrator, it is a new department to be integrated into the hospital's operation. But most important, for the patient, it is better health care.

Engineers interfacing with the medical profession are encouraged to learn as much as possible about medical and hospital practice and, in particular, about the physiology of the human body. It is only by gaining such an understanding that they can communicate intelligently with the members of the medical profession.

1.2 What is Medical Electronics?

Medicine is not new, electronics is not new; but "medical electronics" is new. For the medical community, medical electronics is an extension of the physician's ability to diagnose an ailment and treat it. To put it in another way, let us consider what a doctor could do without the use of electricity. He could do some things, such as: give anesthesia, measure blood pressure, make incisions, tie off blood vessels, scrape tissues, and other things that can be done mechanically. With electricity, however, major advances are possible: the physician can study a patient before surgery using X-rays and monitors which display physiological parameters; make blood analyses; and can perform complex and exacting surgical procedures (*e.g.*, spot-weld a cornea transplant, observe and repair the stomach or the bladder from the outside with a luminous light, repair a malfunctioning heart, or bombard a tumor with a high level energy, such as the laser). Electronics, therefore, supplies tools which are aids for treating patients. If anything, electronics in medicine has enabled the medical community to enlarge its operation because many more procedures are now possible. Consider what a doctor could do 75 years ago when a patient complained of a pain in his chest and what he can do today!

Even the use of computers will greatly benefit the patient and still not reduce the number of medical personnel required to meet the health service needs of our society. Computers will provide a more comprehensive presentation of data to aid the physician, not replace him.

1.3 What Medical Electronics is Not

Medical electronics is not a replacement of the doctor, the nurse, or the lab technician. Just as computers did not throw thousands of people out of work, medical electronics will not put medical professionals on welfare. As a car cannot go on forever without lubrication, so too, electromedical equipment cannot operate without periodic maintenance and overhaul. Equipment can be dropped or otherwise be abused or misused and require repair. Thirdly, medical electronics cannot administer an injection of serum; it cannot tie a suture; it cannot aim the X-ray; it cannot set the broken bone; it cannot swallow a pill for a patient; it cannot prevent a person from tripping and breaking a rib in the first place. It cannot cure the common cold.

1.4 Man's Hierarchy of Organization

The human body is a source of numerous signals, highly significant for diagnosis and therapy. Its internal communication system, provided by the nervous system, is based on the transmission of electrical impulses which results in the co-ordinated activities of the human body. Within the human body can be found electrical, mechanical, thermal, hydraulic, pneumatic, chemical, and various other types of systems, each of which communicates with an external environment, and internally with the other systems of the body. By means of a multilevel control system and communications network, these individual systems are organized to perform many complex functions. Through the integrated operation of all these systems, and their various subsystems, man is able to sustain life, learn to perform useful tasks, acquire personality and behavioural traits, and even reproduce himself.

Measurements can be made at various levels of man's hierarchy of organization. For example, the human being as a whole (the highest level of organization) communicates with his environment in many ways. These methods of communicating could be regarded as the inputs and outputs of the black box and are illustrated in Figure 1.1. In addition, these various inputs and outputs can be measured and analyzed in a variety of ways. Most are readily accessible for measurement, but some, such as speech, behaviour and appearance, are difficult to analyze and interpret.

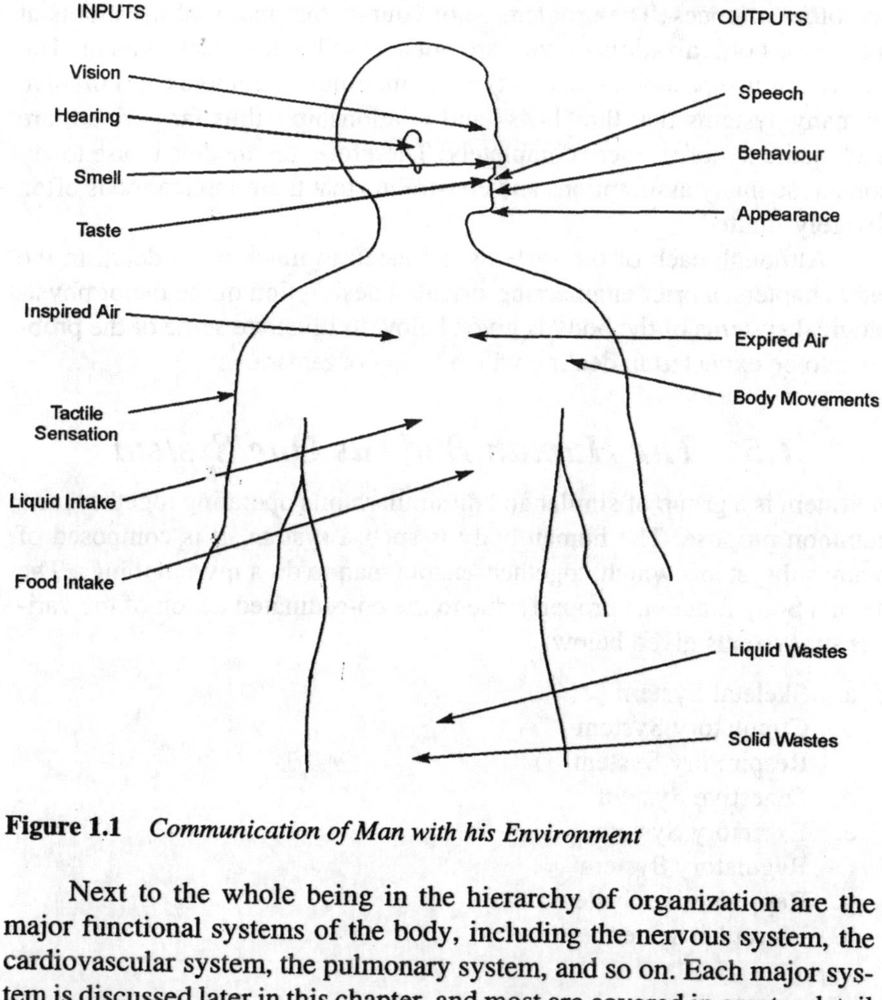

Figure 1.1 *Communication of Man with his Environment*

Next to the whole being in the hierarchy of organization are the major functional systems of the body, including the nervous system, the cardiovascular system, the pulmonary system, and so on. Each major system is discussed later in this chapter, and most are covered in greater detail in later chapters. Just as the whole person communicates with his environment, these major systems communicate with each other as well as with the external environment.

These functional systems can be broken down into subsystems and organs, which can be subdivided into smaller and smaller units. The process can continue down to the cellular level and perhaps even to the molecular level. The major goal of biomedical instrumentation is to make possible the measurement of information communicated by these various elements. If all the variables at all levels of the organizational hierarchy could be measured, and all their interrelationships determined, the functions of the mind and body of man would be much more clearly understood and probably be completely defined by the presently known laws of physics, chemistry, and other sciences. The problem is, of course, that many of the inputs at the various organizational levels are not accessible for measurement. The interrelationships among elements are sometimes so complex and involve so many systems that the "laws" and relationships, thus far derived, are inadequate to define them completely. Therefore, the models in use today contain so many assumptions and constraints that their application is often severely limited.

Although each of the systems is treated in much more detail in the later chapters, a brief engineering-oriented description of the major physiological systems of the body is given below to illustrate some of the problems to be expected in dealing with a living organism.

1.5 The Human Body as One System

A system is a group of similar and dissimilar units operating together for a common purpose. The human body is such a system. It is composed of many subsystems which, together, enable man to do a myriad things. The human body functions properly due to the co-ordinated action of the various subsystems given below:

a. Skeletal System
b. Circulatory System
c. Respiratory System
d. Digestive System
e. Excretory System
f. Regulatory System
g. Reproductive System
h. Muscular System

a. Skeletal System

If we assume the human body to be a building, its framework is the skeleton. The main purpose of the skeletal system is to provide mechanical stability for the body, to protect the delicate organs inside and to serve as an anchorage for the muscles in order to make movement possible through a lever action. The skeleton also serves as a reservoir for calcium and phosphorus and contains the bone marrow in which blood cells are formed. The skeleton consists of 206 bones. The skeletal bones are completely enclosed by a membrane called the *periosteum*, from which a new bone is formed in the healing of fractures. At the top of the skeleton, the *cranium skull* is placed. At the middle, we have the *pelvis*. At the bottom, we have the *foot bones*. Friction in the joints is reduced by a smooth articular cartilage that covers the bone surface. These surfaces, along with the articular capsule attached to the bones, form a closed cavity. The inner surface of the articular capsule is covered by a membrane that secretes a viscous, lubricating fluid.

b. Circulatory System

The circulatory system is a type of transport system. In this system, the blood is the carrier for the transportation of nourishments and waste products within the body. These include, in the former, oxygen, water, electrolytes, hormones, special secretions from glands, and nourishments; and in the latter, carbon dioxide and waste products. The pump for this system is the heart, through which all the blood flows. Blood flows into the heart from the veins which have collected waste products as a result of cell metabolism (see Chapter 2). The blood is then pumped from the heart up to the lungs, where carbon dioxide is diffused out and oxygen diffused in. The blood, now rich in oxygen, returns to the heart from where it is then pumped out to the various organs of the body to meet the cellular needs. Many actions must occur in a very timely order for this flow to be continuous. The heart beats (pumps) at an average rate of about 72 beats per minute in a normal adult, although this figure can vary widely.

c. Respiratory System

The respiratory system is concerned with breathing and respiration. Its main organs are the lungs. The respiratory system starts from the nose and ends in the lungs. We breathe through our nose. The nose has a cavity which leads to the wind-pipe, known as the *nasal cavity*. The nasal cavity contains some hair at the opening which filters dirt particles as the air passes through the nose. The nasal cavity is also wet at the back. So when we breathe through the nose, we inhale air that is free from dirt. Breathing in is called *inspiration* and breathing out is called *expiration*. The air passes from the nose to the wind-pipe and from the wind-pipe to the two bronchi, where it is distributed to the left and right lungs. In the lungs, the air stream is distributed throughout the bronchial tree via the finer airtubes to the *alveoli*, where the gas exchange takes place.

d. Digestive System

The digestive system is responsible for digesting the food one eats. In the digestive system, the food is disintegrated chemically so that it can be absorbed. First, the food is cut into smaller pieces and mixed with saliva in the mouth. The gullet carries the food from the mouth to the stomach. The stomach wall contains various kinds of glandular cells which secrete hydrochloric acid, pepsin, and mucus. The food remains in the stomach for about four to five hours. Here, it is made into a paste by the juices of the stomach. The bile is produced by the liver. Then the food passes to the small intestine, where it is further digested. The small intestine is connected to the large intestine. It keeps the food long enough to absorb the water contained in the waste. The waste is thrown out of the body as faeces. Thus, the vital processes require energy which is obtained through metabolism, whereby the digestive system burns carbohydrates, fats and proteins. In the combustion process, oxygen is consumed; it is taken up by the respiratory organs. The combustion products consist of carbon dioxide, water, urea and other nitrogenous substances. The combustion products are excreted by the respiratory system and the excretory system. Thus,

$$\text{Food} + \text{Oxygen} \xrightarrow{\text{Metabolic Activity}} \text{Energy} + \text{Wastes (Carbon Dioxide)}$$

The digestive system is in the form of a long tube consisting of the mouth at its upper end and the anus at its lower orifice and the food is advanced by peristaltic movements produced by automatic rhythmic contractions of the smooth muscles, present throughout the wall of the digestive tract.

e. Excretory System

The excretory system removes the waste products, formed during the combustion of the food, from our body. The lungs, the kidneys, the skin and the large intestine form the excretory system. The lungs remove carbon dioxide. The nitrogenous breakdown products formed during the combustion of proteins are excreted through the kidneys in the form of urine. Thus, the excretion of most of the volatile substances occurs through the lungs and of most of the non-volatile substances through the kidneys. The sweat glands in the skin remove water, salt and other waste products. The large intestine removes the undigested food through the anus. Further, the liver, via the bile, removes certain waste products. There is also some excretion through the colon.

f. Regulatory System

The regulatory system can be divided into two types—the *Nervous system* for the regulation of rapid events and the *Endocrine system* for the regulation of slower metabolic processes. The nervous system governs the rapid events like muscular contractions and secretions of most of the glands in the body. It ensures the smooth functions of various systems in the body. It works through a system of nerves which act as tiny electric wires. *Sensory nerves* carry the information gathered by the sensory organs to the brain. Thus, they serve as the message carriers for the brain. *Motor nerves* carry back the order from the brain to the muscles and glands. Mixed nerves perform the functions of both sensory and motor nerves, i.e., they carry signals or impulses from the sensory organs to the brain; at the same time, they convey the order of the brain or the spinal cord to the other parts of the body. Thus, the nervous system co-ordinates the functions of organs. This occurs by the action of external and internal stimuli which result in reflexes. The nervous system consists of the *central nervous system* (CNS) and the *peripheral nervous system*. The central nervous system is made up of the brain and the spinal cord. The peripheral nervous system consists of all the nerves and groups of neurons (nerve cells) outside the brain and spinal cord.

The endocrine system works by using hormones which are carried through the circulatory system. The hormones are generated in the endocrine glands. The principal endocrine gland, the *pituitary gland*, governs several other endocrine glands. The *thyroid gland* secretes thyroxine which increases the metabolism in the body. The deficiency of the thyroid hormone results in low metabolic rate and mental retardation. The *adrenal gland* secretes corticoids and they regulate the blood pressure and flow. The *pancreas gland* secretes insulin which is a hormone that regulates the metabolism of glucose. A deficiency of insulin results in increased glucose in the blood, which leads to the condition called diabetes. The *duodenal gland* produces hydrochloric acid using its hormones. The *gonads*, the sex glands, generate sex hormones (testosterone for males and estrogen and progesterone for females). These hormones regulate sex drive and pregnancy. The *placenta* also generates a number of hormones which are useful during pregnancy.

g. *Reproductive System*

A fetus develops through repeated cell division by means of the reproductive system. The seminal vesicles and prostate gland in the male genital organs produce two secretions which form the *semen*. This semen activates the sperm and makes it mobile. The ovaries in the female genital organ produce the *ova* and *eggs*. In ovulation, usually one egg is detached and spends a few days passing though the fallopian tube to the uterus. Fertilization takes place in the fallopian tube to which the sperm have swum through the uterus. When the ovum reaches the uterus, it becomes embedded in the endometrium where the fetus is developed through repeated cell division. The fetus receives nourishment from the mother via the umbilical cord and the placenta which is attached to the uterine wall.

h. Muscular System

The movements of the various parts of the body are caused by the muscles. There are three kinds of muscles: *voluntary* (skeletal) *muscles* which work at our will (example: arm muscles), *involuntary* (smooth) *muscles*, which work without our knowledge (example: muscles in the food canal), and *cardiac muscles* which help in functioning of the heart, and work day and night without any rest. We know that the contraction of the muscles is regulated via the nerves. There are two types of contractions: isotonic and isometric. In an isotonic contraction, the muscle is shortened under constant load and in an isometric contraction the muscle contracts without shortening. In a muscular movement, the first stage is usually an isometric contraction subsequently followed by an isotonic contraction.

Skeletal muscles have been more intensively studied and are better understood than the other two muscles mentioned above. The skeletal muscle system is a means by which a living being reacts to the external environment. The performance of smooth, efficient and co-ordinated bodily movements is the outward sign of complex and extensive activity in the central nervous system. A muscle is composed of many fibres. Each fibre consists of many bundles of closely packed 'fibrils'. Fibrils, in turn, are made up of two types of filaments—*thick* and *thin,* which so interact that they slide past each other during shortening.

Muscular contraction is probably the most impressive example of "living machinery".

Chapter 2

The Human Cell—Action Potential

2.1 Cells and their Structure

The basic building block of the human body is the cell. This special combination of molecules, acting and reacting in a certain way and being able to reproduce itself, constitutes the smallest level of human life as we know it. The average adult is said to consist of about 100 trillion such cells, with about half of his/her body being fluid in character. Because not all cells are identical, special combinations of cells have evolved forming different functioning structures. All the cells in one body, however, have united, in a sense, to form the total body.

The cells of the same general function are grouped together to form tissues, such as muscle nerves, blood, bones and cartilage. More commonly, two or more tissues of dissimilar physiology are combined in a particular pattern to form larger functional units called organs—heart, lungs, kidney, skin, blood vessels, and so on. Several organs whose functions are interrelated constitute a system. For example, the digestive system (the mouth, trachea, stomach, intestines, liver and rectum) allows food to be ingested, breaks it down to the molecular level, allows it to be absorbed into the blood and removes any waste products; the other systems being the circulatory system, respiratory system, nervous system, endocrine system, excretory system, reproductive system, *etc*.

The cells of the tissues are held together by a product formed by the cells themselves, that is called *intercellular cement.* In order to sustain the cells, a fluid, called *interstitial fluid* or *extracellular fluid,* is able to flow to each cell to bring nourishment to it, to remove waste products from it, and to respond to electrochemical impulses. It is interesting to note that about half of the body is water; of this half, three-fourths of the water is in the cells, the other quarter in the spaces between the cells.

2.2 Characteristics of Living Organisms

A living organism (virus, cell, fish, dog or man) must have or display the following attributes or activities.

a. Metabolism

Metabolism is the process whereby a living organism ingests substances containing energy, converts these substances to liberate this energy, uses this energy to maintain activities, and excretes waste products. An organism is able to maintain itself for a period of time because of this process.

b. Motility

Motility is the ability of the organism to move on its own without outside assistance. This action enables the organism to warm itself, secure food, avoid threatening situations as well as perform other tasks.

c. Irritability

Irritability is the ability of the organism to respond to an outside influence ("stimulus"). A stimulus can take many forms, such as a change in temperature, an electric current, or an intrusion of a foreign object or harmful bacteria. The response to a stimulus can also take many forms, such as contraction, expansion, discharge, movement, or chemical and physical changes.

d. Growth

The ability to grow must also be evident. Literally, this means to increase in one or all of the following: size, energy, form, chemical structure and function. This criterion might also be called the "ability to change".

e. Reproducibility

Reproductivity is the ability to reproduce one's own kind or species. This may occur by self-division or by the union of two organisms of the same type but different genders.

These five basic operations must be in evidence for an organism to be considered 'living'. Man, the most complex living organism, displays these criteria because his basic building block, the cell, possesses these criteria.

2.3 General Characteristics of the Human Cell

The "average" human cell can be described as follows (Figure 2.1):

(a) It is nearly spherical

(b) It is about twenty microns in diameter.

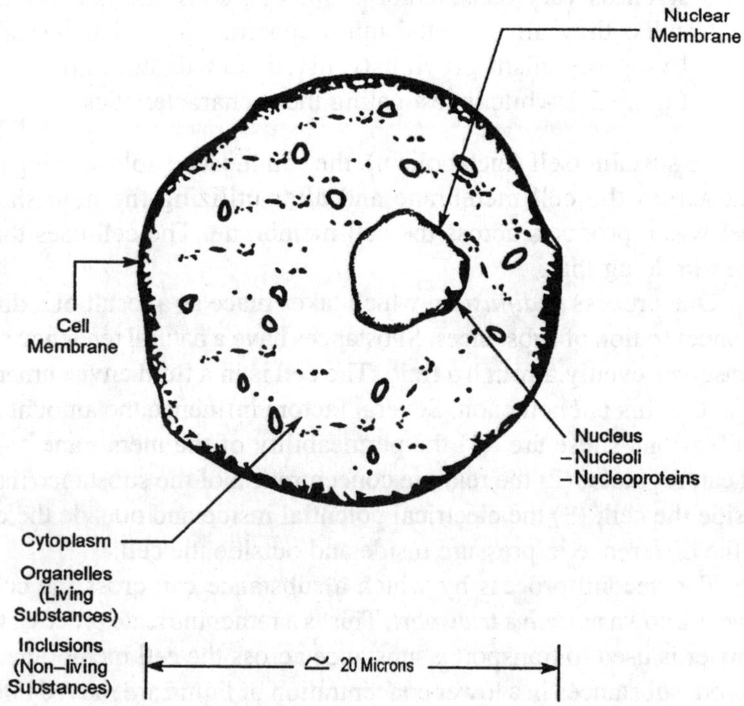

Figure 2.1 *"Average" Human Cell*

(c) It has a semipermeable outer membrane with a thickness of about 100 Angstrom units, through which certain substances can pass.

(d) It has an inner fluid substance called cytoplasm, comprising of organelles and inclusions. Organelles consisting of small internal organs of the cell (living substances); and inclusions made of lifeless accumulations of metabolites, or cell products such as fats and carbohydrates.

(e) It has an inner "control unit" called a *nucleus*, which has a membrane (nuclear membrane) and is filled with a substance called nucleoplasm. Within this nucleus are nucleoli which contains a large amount of Ribonucleic acid (RNA), *genes or chromosomes* containing large quantities of Deoxyribonucleic acid (DNA) which determine the specific biochemical characteristics of each cell type and controls its metabolic activities. The DNA does not occur free in the nucleoplasm but is available in a combined form in the nucleoproteins. The above description is for the "average" human cell, and such things as shape, the number of nuclei, size and responsiveness vary considerably. One characteristic is common to all cells: they all meet the minimum function characteristics of a living organism previously listed. Let us continue to refer to Figure 2.1 while investigating these characteristics.

To sustain itself (metabolism), the ceıı must be able to bring nourishment across the cell membrane and after utilizing the nourishment, to expel waste products across the cell membrane. The cell uses three processes in doing this.

One process is *diffusion*, which takes place as a result of a difference of concentration of substances. Substances have a natural tendency to spread themselves evenly through a fluid. The cell is in a fluid environment and is subject to this phenomenon. Several factors influence the amount and rate of diffusion. These are: (1) the permeability of the membrane, which the cell can regulate; (2) the relative concentration of the substance inside and outside the cell; (3) the electrical potential inside and outside the cell; and (4) the difference in pressure inside and outside the cell.

The second process by which a substance can cross the cell membrane is known as *active transport*. This is a rather intricate process, whereby a carrier is used to transport a substance across the cell membrane. By this method, substances in a lower concentration of liquid are moved into higher levels of concentration. Very little is known about how the cell knows which substances to transport across the membrane. This process differs markedly from diffusion, in that it is a deliberate action which creates an imbalance of charges and energy. Diffusion, on the other hand, is passive, and is a tendency to bring about a balance of charges. Again, the cell has mechanisms which keep track of this balance or imbalance in order to survive.

The third process is similar to active transport, except that a carrier is not used. It is called *pinocytosis*. When certain substances approach and rest on the cell membrane, the cell membrane can "engulf" them, and form a sort of capsule; this capsule then travels through the membrane into the cytoplasm.

2.4 Resting Potential

Any animal or human tissue, such as the muscle or brain, is composed of closely packed cells. The solution surrounding and bathing the cell is known as the interstitial fluid and the one inside is known as the intracellular fluid. A thin, highly-organized membrane separates the two fluids. The concentration of the ions inside and outside the cell are markedly different.

(i) In the interstitial fluid, the concentration of sodium (Na^+) and chloride (Cl^-) ions are much higher than in the intercellular fluid.

(ii) In the intercellular fluid, the concentration of potassium ions (K^+) is much higher than in the interstitial fluid.

(iii) There is an electric potential difference between the inside and outside fluids. In skeletal muscles, the inside of the cell is about 90 mV negative with respect to interstitial fluid.

Figure 2.2(a) is a hypothetical diagram showing the membrane as having pores bathed in interstitial and intercellular fluids. The simplified diagram is an aid to describe the generation of voltage gradients by the diffusion of ions. It is not a portrayal of a real membrane.

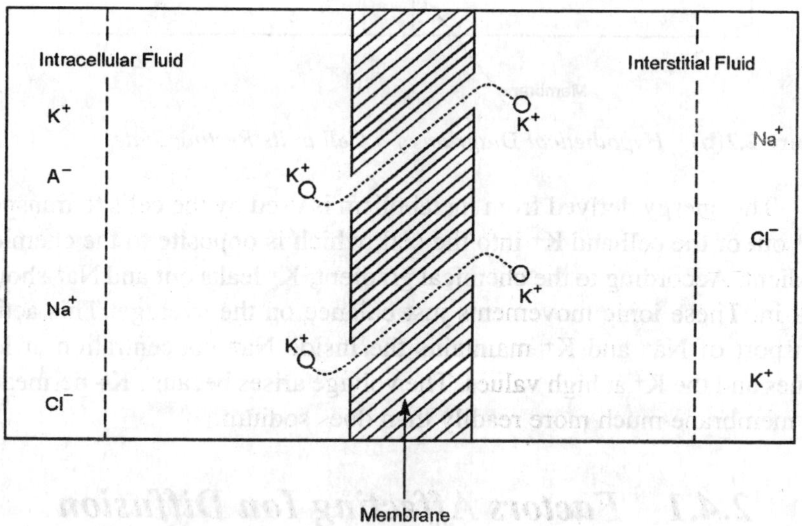

Figure 2.2(a) *Hypothetical Diagram of a Cell showing the Membrane as having Pores. Size of Symbols in Left and Right Hand Columns Indicate Relative Concentration of Ions*

The membrane plays an important role in the maintenance of these differences. In the resting condition, that is, when the membrane is not stimulated, it is highly permeable to potassium but only slightly permeable to sodium. Because of this and because of its steep concentration gradient, potassium tends to leak out at a high rate. The result is that the inside of the cell becomes electrically negative with respect to the outside. However, there is a limit to this process. An equilibrium is reached when the tendency for potassium to diffuse out is balanced by the electric field that has been set up. At this point, the interior of the cell is about 60 to 90 mV negative relative to the external fluid. This difference in potential between the outside and inside of the cell is called the *resting potential of the cell.*

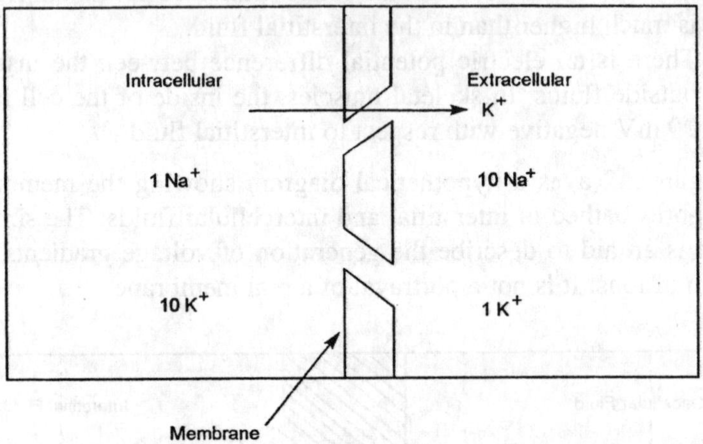

Figure 2.2(b) *Hypothetical Diagram of a Cell in its Resting State*

The energy derived from metabolism is used by the cells to transport Na^+ out of the cell and K^+ into the cell, which is opposite to the chemical gradient. According to the chemical gradient, K^+ leaks out and Na^+ should leak in. These ionic movements just balance on the average. This active transport of Na^+ and K^+ maintains the inside Na^+ concentration at low values and the K^+ at high values. The voltage arises because K+ permeates the membrane much more readily than does sodium.

2.4.1 Factors Affecting Ion Diffusion through Membranes

(i) The outside fluid has a high concentration of Na^+ and Cl^-.
(ii) The inside fluid has a high concentration of K^+ and the organic anions (A^-).

Passive Factors

(i) Due to the concentration gradient, Na^+ and Cl^- tend to diffuse into the cell and K^+ and A^- tend to diffuse out of the cell.
(ii) The rate of diffusion of these substances not only depends on the concentration difference, but also on the permeability of the membrane separating the inside and outside of the cell to these substances.

2.5 Nernst Equation

As discussed earlier, if the membrane permeability to one ion is greater than to another and if ion concentration differences exist across the membrane, then a transmembrane voltage will develop.

In the hypothetical diagram of a cell shown above (Figure 2.2(b)), the permeability of the membrane to potassium ions K^+, (P_K) is very high compared to that of sodium ions Na^+, (P_{Na}) under resting conditions. K^+, therefore, diffuses out due to the concentration gradient, and as $P_K > P_{Na}$, this generates an electric field which retards the outward diffusion of K^+. Finally, V (transmembrane potential), reaches a stable value when K^+ efflux equals K^+ influx.

The transmembrane potential which equalizes fluxes for a particular ion is called the *equilibrium potential for that ion*. The potassium equilibrium potential can be written as:

$$V_K = 61.6 \log_{10}\left[(K^+)_o/(K^+)_i\right] \text{mV} \quad \text{...(2.1)}$$

(o stands for outside and i stands for inside the cell).

This is the NERNST Equation which can be derived in the following manner:

(i) In order to move an ion in a solution in which potential difference is present, work must be done with or against two factors: (a) the electric field, (b) any concentration differences in the ion. Thus, both electrical and concentration work must be performed.

(ii) The electrical work, in joules, necessary to move a mole of ion against a potential difference V in volts is given by the expression

$$\text{Work} = FV \quad \text{....(2.2)}$$

where F is the Faraday constant (number of coulombs per mole of charge = 96,500).

(iii) The work required to move a mole of ion from a concentration C_1 to a concentration C_2, where C_2 is higher than C_1, is given by the expression,

$$2.303\ RT \log_{10}(C_2/C_1) \quad \text{....(2.3)}$$

where R is the gas constant and T the absolute temperature. If the ion is moving with the gradient, the energy is obtained from the concentration difference.

(iv) The total work is given by the sum of the electrical and concentration work

or $$\text{Work} = FV + 2.303\ RT \log_{10}(C_2/C_1) \quad \text{....(2.4)}$$

(v) When the system is in equilibrium, the total work required is zero. So,

$$FV = -2.303\ RT \log_{10}(C_2 / C_1) \quad \text{....(2.5)}$$

or $$V = 2.303\ RT/F \log10(C2/C1) \quad \text{....(2.6)}$$

substituting R = 8.31 joules per mole-degree abs, T = 273 + 37 degrees abs (37°C being the normal, healthy human temperature), F = 96,500 coulombs/mole,

$$V_K (K^+ \text{ equilibrium potential}) = 61.6 \log_{10} [(K^+)_o/(K^+)_i] \text{ mV} \quad \text{...(2.7)}$$

This is the *Nernst Equation* and can be written for every ion present in the system.

The physical interpretation of this equation is that, the tendency of an ion to diffuse down its concentration gradient is countered by the building up of an electric field at the junction of the two solutions. The direction of the field is such as to hold back the ion from further movement. It is to be noted that it is only necessary to move a very minute number of ions in order to produce the restraint required to obtain the equilibrium of electric and concentration gradients.

2.6 Electrical Characteristics of the Human Cell

The structure and chemical composition of the human cell is such that a voltage difference exists between compounds inside and outside the cell. Further, the cell is composed of some compounds that are electrically conductive and will cause a momentary current to flow (an *action potential*) when the cell is stimulated. Stimulation includes electrical, chemical, and mechanical changes as well as changes in temperature. This action potential can occur because of the "semipermeability" of the cell membrane.

Semipermeable means that some substances can pass through a material while others cannot. In cells, certain substances are ingested (by one of the three processes mentioned previously) through what are thought to be special porous holes in the *cell membrane*; other substances are kept out. Conversely, waste products are passed out while other substances are kept in. Many factors determine how permeable the membrane will be at any particular moment. With regard to two elements, sodium ion (Na^+) and potassium (K^+), it has been found that, with the cell in its resting state, the sodium concentration outside the cell is about ten times that of the concentration inside the cell; and that the potassium concentration is almost the reverse. Further, if a voltmeter is positioned as shown in Figure 2.3, a reading of minus 90 millivolts will be observed on the meter. This difference in concentration of each type of ion (sodium and potassium being the most influential, though there are about fifteen other ions in the fluid inside and outside the cell) results in the potential difference existing between the outside and the inside of the cell. The value of this voltage for the resting cell will depend on: (a) the relative difference in concentration (the *chemical gradient*), and (b) any external sources resulting from the transport of ions across the membrane (the *electrical gradient*).

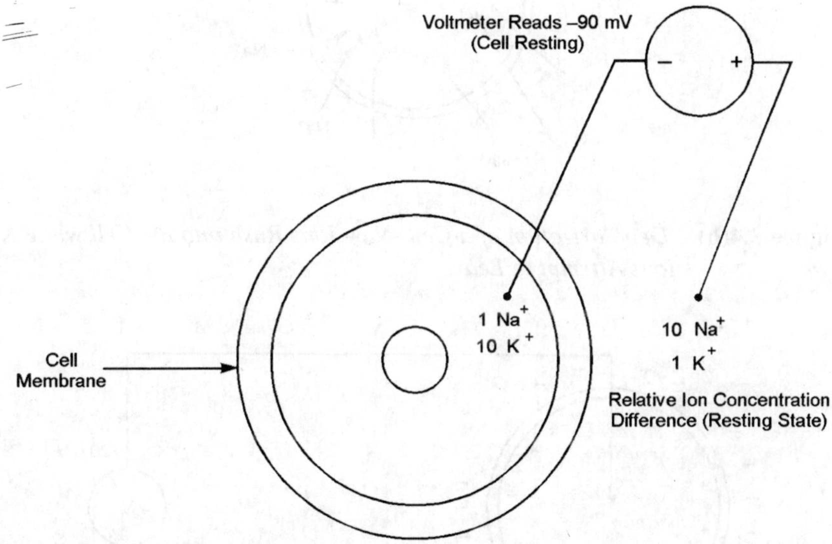

Figure 2.3 *Resting Cell*

When a cell receives a stimulus above a certain "threshold" value, this balance is upset, and the cell will go through a cycle known as an "action potential". It is by this method that impulses, technically called *electrochemical impulses,* are transmitted from cell to cell. It is on this

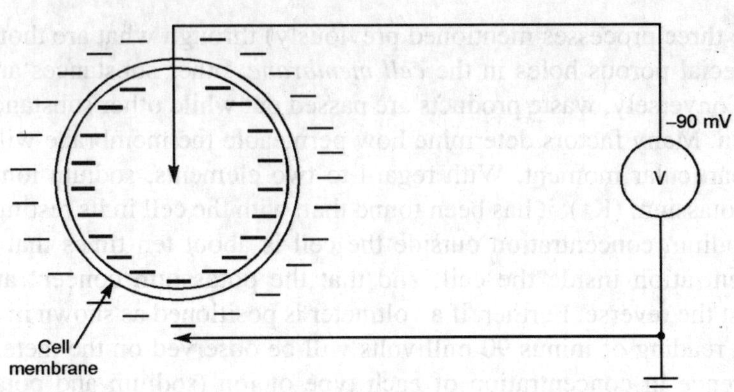

Figure 2.4(a) *Polarized Cell with its Resting Potential*

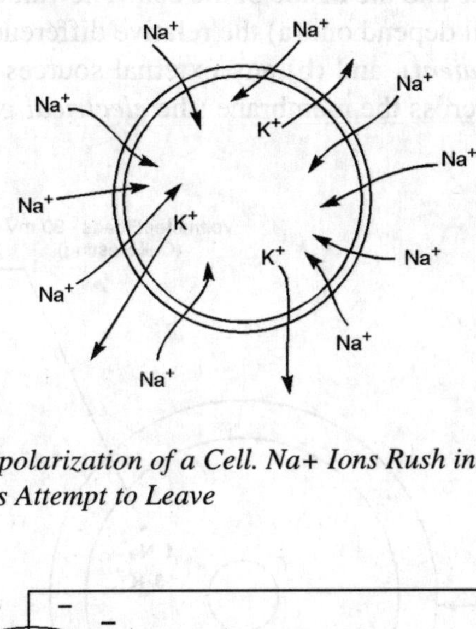

Figure 2.4(b) *Depolarization of a Cell. Na+ Ions Rush into the Cell while K^+ Ions Attempt to Leave*

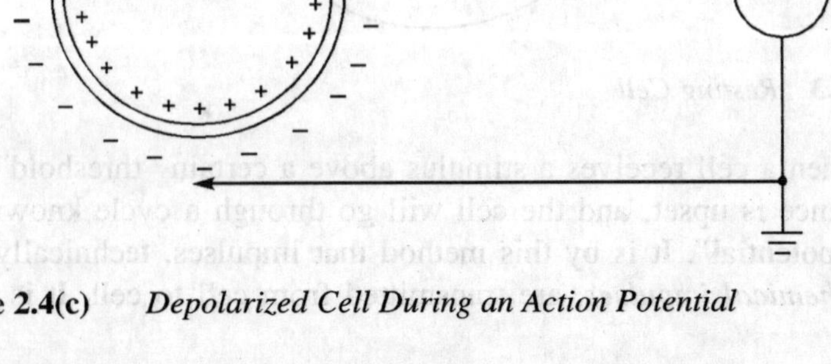

Figure 2.4(c) *Depolarized Cell During an Action Potential*

basis that "gross level" actions can occur—the co-ordinated action potential of millions of cells. The action potential results from a change in permeability of the membrane at the point of stimulation, with a spreading effect following.

Here is the sequence of events of an action potential after a stimulus. Initially, in its resting state, a cell has a negative charge along the inner surface of its membrane and a lesser negative charge along the outer portion (Figure 2.4(a)). The cell, in such a condition, is said to be polarized and in effect acts as a tiny biological battery. Experiments showed that the internal resting potential within cell is approximately –90 mV with reference to the outside of the cell. When excited or stimulated, the membrane permeability to sodium (Na^+) ions is increased, resulting in sodium ions flowing into the cell (Figure 2.4(b)), thereby, the outer side of the cell membrane becomes momentarily negative with respect to the interior (Figure 2.4(c)). The voltage inside the cell rises (becomes less negative). This process is called *depolarization* and the cell potential changes to approximately +20 mV. At this stage, when the electrical balance is achieved again, the cell is referred to as being in a depolarized state. The cell membrane then begins to return to its original permeability, beginning at the point of stimulation and spreading to the rest of the membrane. Rather than sodium ions flowing out, however, potassium ions diffuse out, causing the cell to return to its normal resting potential of –90 mV, and the *repolarization* of the cell is said to have occurred. The discharging and recharging of the cell produces the voltage wave which can be recorded by suitable methods using microelectrodes. A typical cell "action potential" waveform, so recorded, is shown in Figure 2.5. The time-scale for the action potential depends on the type of cell producing the potential. For example, it is of the order of 1 ms total duration in nerve and muscle cells and 150 to 300 ms in the heart muscle cells.

It should be noted that the currents involved in bioelectricity are unlike the currents involved in *electronics*. Bioelectronic currents are due to positive and negative ion movement within a conductive fluid. As these ions possess finite mass and encounter resistance to movement within the fluid, their speeds are limited. The cell action potential, thus, shows a "finite rise time" and "fall time".

The ionic concentration gradient across the cell membrane is maintained by virtue of the metabolic energy expended by the cell in "pumping" ions against the ionic gradient formed by the differing ionic concentrations between the inside and outside of the cell. This action has been referred to as the *Sodium Potassium Pump*. Although little is known of the exact chemical steps involved in the sodium pump, it is generally believed that sodium is withdrawn against both the charge and concentration gradients,

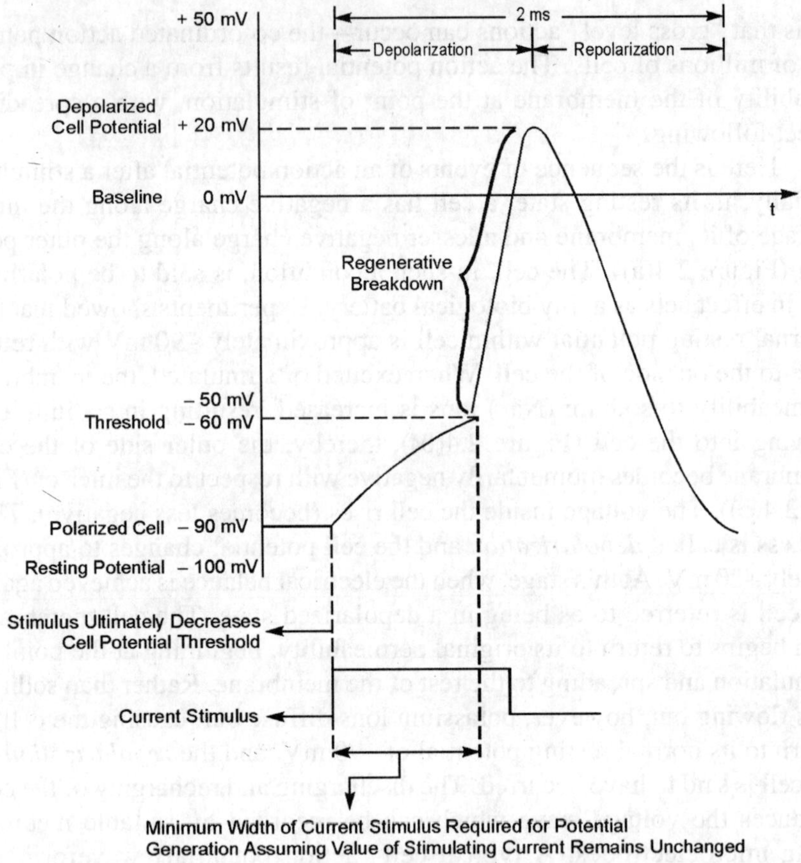

Figure 2.5 *Cell Action Potential (internally recorded with a microelectrode)*

supported by some form of high-energy phosphate compound. The rate of pumping is directly proportional to the sodium concentration in the cell. It is also believed that the operation of this pump is linked with the influx of potassium into the cell, as if a cyclic process involving an exchange of sodium for potassium exists.

A cell may be stimulated, or caused to depolarize and then repolarize, by subjecting the cell membrane to an ionic current. However, unless a stimulus above a certain minimum value is applied, the cell will not be depolarized and no action potential is generated. This value is known as the stimulus threshold; typically, this voltage is about –60 mV. The stimulus threshold is a function of both current and time; the threshold may be exceeded by a short, high current pulse or by a longer, lower current pulse.

Regardless of the method of excitation of cells or the intensity of the stimulus, which is assumed to be greater than the threshold of stimulus, the action potential is always the same for any given cell. This is known as the *all-or-nothing law*. A stimulus voltage generally does not affect a cell while it is changing its polarization. The net height of the action potential is defined as the difference between the potential of the depolarized membrane at the peak of the action potential and the resting potential.

Following the generation of an action potential, there is a brief period of time during which the cell cannot respond to any new stimulus. This period is called the *absolute refractory period* and lasts about 1 ms in nerve cells. Following the absolute refractory period, there occurs a *relative refractory period*, during which another action potential can be triggered, but a much stronger stimulation is required. In nerve cells, the relative refractory period lasts several milliseconds. These refractory periods are believed to be the result of after-potentials that follow an action potential.

Some characteristics of the action potential are:

(a) Once the "threshold" has been reached, the sequence cannot be turned off or changed, regardless of how long the stimulus is continued or if it is removed.

(b) A second stimulus will not have any effect while the cell is in the depolarizing phase.

(c) When a cell is repolarizing, a greater than normal stimulus is required in order to initiate another action potential.

(d) The threshold level can be increased or decreased by altering the permeability of the membrane, by changing the concentration levels of sodium, potassium, calcium, *etc.*; or by introducing drugs into the fluid surrounding the cell.

(e) Once the threshold is reached, the action potential will take about 10 ms to depolarize the repolarized cells.

Methods have been devised, using very small probes and stimulators, for viewing the action potential on an oscilloscope. The tracing will resemble that in Figure 2.5 depending on the type of the cell, its condition and its environment.

2.7 Bioelectric Potentials—Origin

The depolarization and repolarization undergone by a cell from time to time, give rise to voltage waveforms, which are of interest to clinicians and biomedical engineers. Thus, each cell in the human body is a minute voltage generator, and is the basic source of all *bioelectric potentials*. These bioelectric potentials are actually ionic voltages produced by the co-ordinated electrochemical activity of large groups of cells. In this type of synchronized action of many cells, the charges tend to migrate through the body fluids towards the still unexcited cell areas. Such charge migration constitutes an electrical current and, hence, sets up potential differences between various portions of the body including its outer surface. These potential differences can be conveniently picked up by placing conducting plates (electrodes) at any two points on the surface of the body and measured with the help of a sensitive instrument. The biopotentials, so monitored, are highly significant for diagnosis and therapy. Table 2.1 shows typical bioelectric signals and their characteristics.

Table 2.1 Characteristics of the various Bioelectric Signals

Bioelectric Signal	*Spectrum* (Hz)	*Potential Range* (μV)	*Sensing Devices Used*	*Signal Origin*
Electrocardiogram (ECG) Vectorcardiogram	0.05 to 100	10 to 5 000 covers fetal range	Surface electrodes are used with jelly. Needle electrodes are less noisy	Heart muscles
Electroencephalogram (EEG)	0.1 to 100	2 to 200	Surface and needle electrodes	Neuronal activity of brain
Cerebral Potentials (intracranially recorded)	Pulse duration (0.6 ms to 0.1s)	10 to 10 0000	Deep needle electrodes	Cerebrum of the brain
Electromyogram (EMG) (Primary signal)	5 to 2 000	20 to 5 000	Surface or needle electrodes	Skin muscles
Electrogastrogram (EGG)	0.05 to 0.2	10 to 350	Surface electrodes	Peristaltic movement of the gastro-intestinal tract
Electroretinogram (ERG)	0.01 to 200	0.5 to 1 000	Corneal electrodes	Retina of the eye
Electrooculogram (EOG)	DC to 100	10 to 3 500	Miniature surface electrodes	Corneal-retinal potential variations

Chapter 3

Basis of Bioelectric Potentials

A. Cardiovascular System—Electrocardiogram

Coronary heart disease is the single greatest cause of death in the world today. The most frequent reason for death is acute myocardial infarction. Thus, a major percentage of medical effort, both in the research and clinical fields, is directed towards the understanding of the cardiovascular system and the subsequent prevention of heart disease.

3.1 Blood Circulation

In this chapter the cardiovascular system is discussed, not only from the point of view of basic physiology but also with the idea that it is an engineering system. The major component of the circulatory system is the heart which supplies the power required to circulate blood throughout the body. The heart is two pumps in series; the smaller right-hand section provides the power required to force blood through the lungs, and the larger, more powerful left-hand section provides the power required to force blood through the body. A simplified block diagram is shown in Figure 3.1.

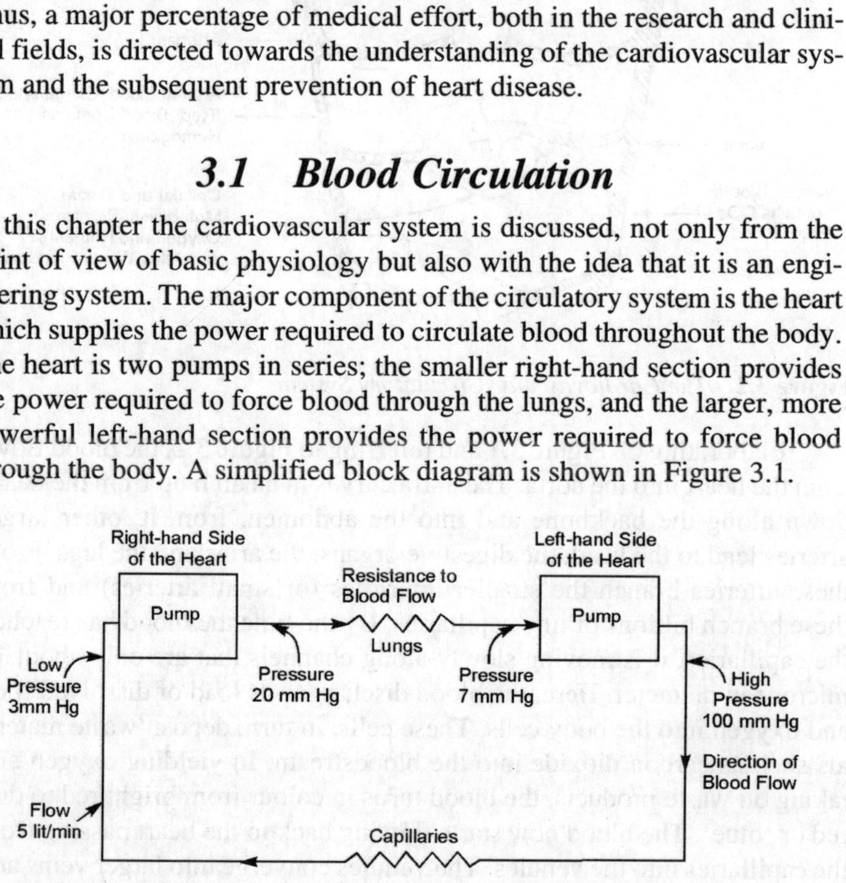

Figure 3.1 *Simplified Block Diagram of the Circulatory System*

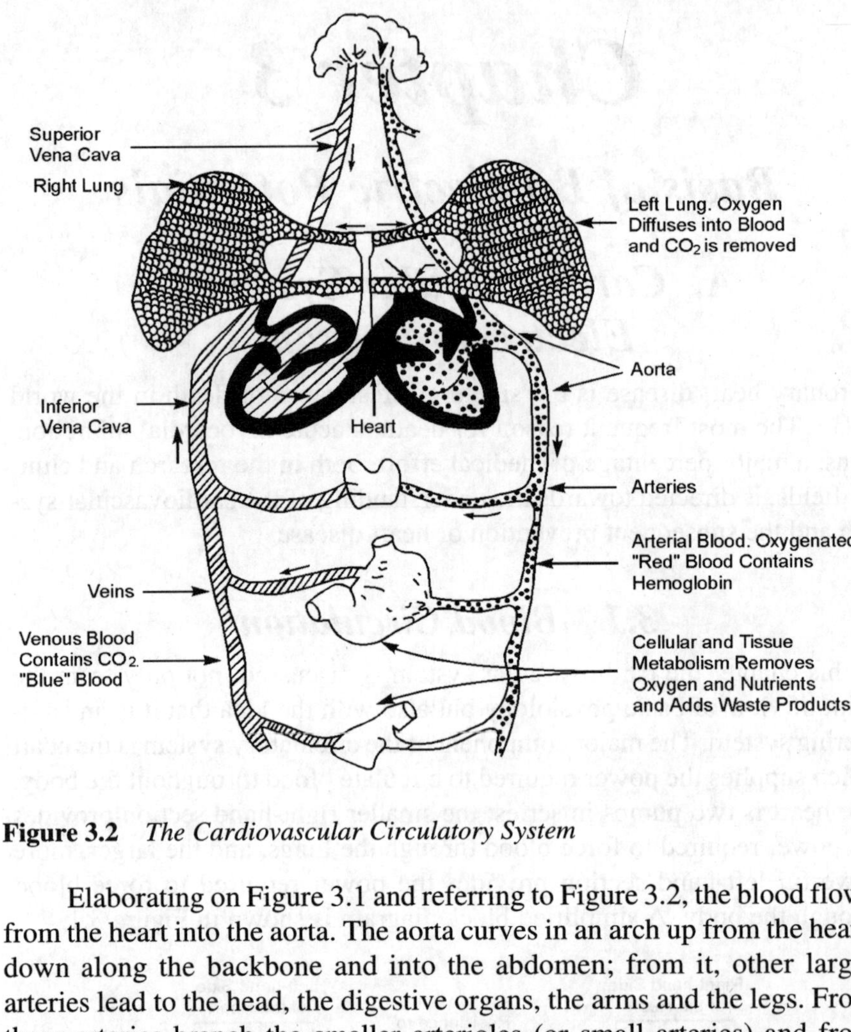

Figure 3.2 *The Cardiovascular Circulatory System*

Elaborating on Figure 3.1 and referring to Figure 3.2, the blood flows from the heart into the aorta. The aorta curves in an arch up from the heart, down along the backbone and into the abdomen; from it, other larger arteries lead to the head, the digestive organs, the arms and the legs. From these arteries branch the smaller arterioles (or small arteries) and from these branch billions of tiny capillaries. By the time the blood has reached the capillaries, it is moving slowly along channels that are only about 10 microns in diameter. Here, the blood discharges its load of dissolved food and oxygen into the body cells. These cells, in turn, deposit waste materials such as carbon dioxide into the bloodstream. In yielding oxygen and taking on waste products, the blood turns in colour from bright red to dull red or 'blue'. The blood now starts flowing back to the heart passing from the capillaries into the venules. The venules converge into larger veins and then into the two largest veins just above and below the heart known as the vena cava. The blood empties into the right atrium, is pumped into the right ventricle through the tricuspid valve, and then moves out through the pulmonary artery, opening the pulmonary valve to the lungs. The lungs supply the blood with fresh oxygen. The blood passes from the lungs to the left atrium, and then is pumped into the left ventricle through the mitral or bicuspid valve, and passes out via the aorta, opening the aortic valve to begin the circulation process again. This general flow throughout the body is known as the "systemic circulation"; the flow to and from the lungs is known as the "pulmonary circulation". Local circulation within the systemic system include the renal (to the kidneys), the hepatic portal (to the liver), the cerebral (to the brain), and the coronary (to the heart itself). The waste products contained in the blood are removed by the kidneys and liver (see Figure 3.2). The average quantity of blood in a man's body is about five litres and is completely circulated through the body in approximately one minute.

3.2 The Heart and its Mechanical Activity

The heart itself weighs less than 450 g, is about 15 cm long at its maximum dimension and lies pointed downward in the chest cavity to the left of the mid-center body line. It is about the size of one's own fist. The heart has a covering as well as lining. Its covering, the pericardium, consists of two layers of fibrous tissues with a small space in between, filled with a thin film of pericardial fluid. The left and right sides of the heart (Figure 3.3) are separated by the septum, or dividing wall of tissue. The walls of the heart are made entirely of muscle; within these walls are four hollow chambers, a left and a right receiving chamber (atrium) and below them a left and a right pumping chamber (ventricle).

The cardiac cycle or the mechanical activity of the heart is characterized by the following events. With a person in a sitting position, the heart beats (or contracts) about 70 times per minute. With each beat, a quantity of blood is driven through the heart. Between the beats, the heart mechanically rests and this is known as the period of "diastole". During diastole, the heart assumes its maximum size and fills with oxygenated blood returning from the lungs and the venous blood returning from the body. The heart's period of mechanical activity is known as "systole". The onset of systole is initiated by the contraction of the muscles surrounding the atria which propel additional blood into the ventricles. The ventricles then begin to contract, thereby causing a rise in pressure in the ventricles. This increased pressure shuts the atrioventricular valves (tricuspid and bicuspid) and, with further contraction, the pressure continues to rise. Once the

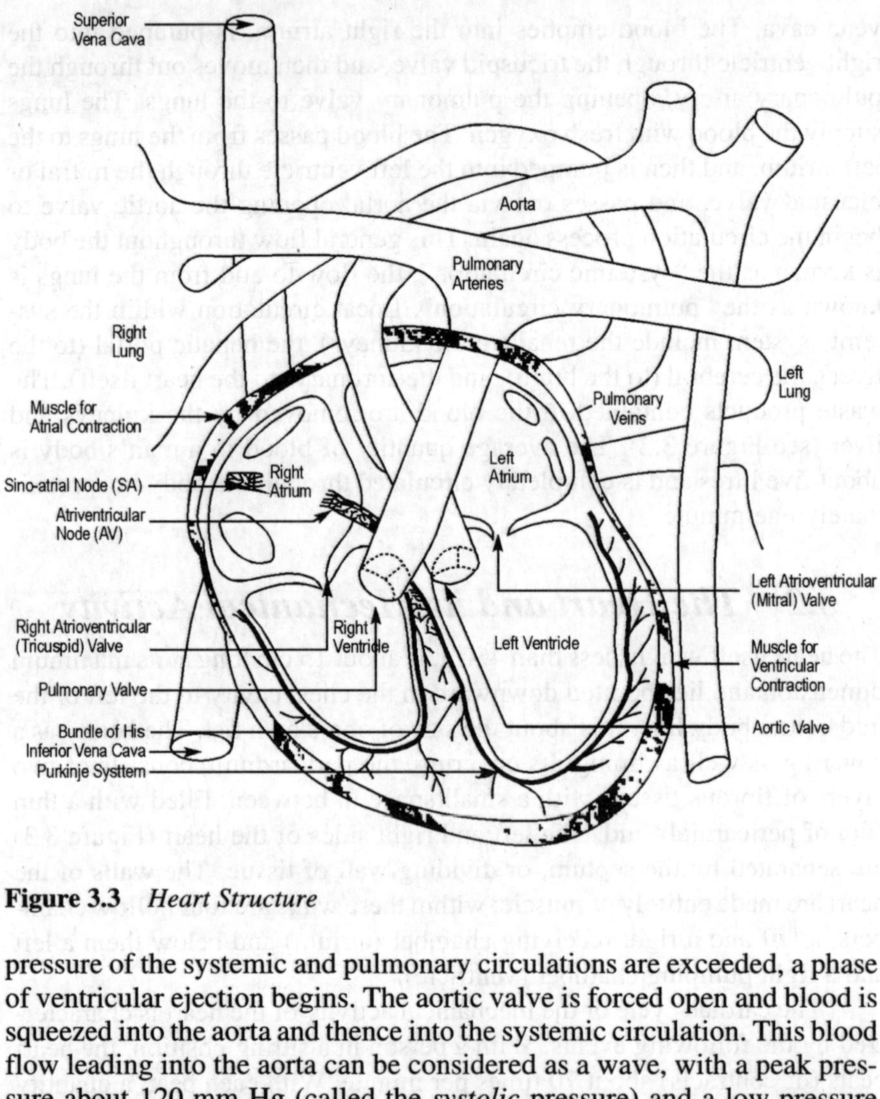

Figure 3.3 *Heart Structure*

pressure of the systemic and pulmonary circulations are exceeded, a phase of ventricular ejection begins. The aortic valve is forced open and blood is squeezed into the aorta and thence into the systemic circulation. This blood flow leading into the aorta can be considered as a wave, with a peak pressure about 120 mm Hg (called the *systolic* pressure) and a low pressure about 80 mm Hg (called the *diastolic* pressure). There is a slight back pressure built up as the mitral valve closes, and this shows up as the *dicrotic notch* in Figure 3.4. Likewise, the pulmonary valve is forced open and blood is supplied to the pulmonary circulation. After the ventricular contents are partially ejected, the muscles surrounding the ventricles relax and the ventricular pressures fall. As soon as these pressures fall below the pressures sustained in the circulatory system, the aortic and pulmonary valves close, signalling the onset of diastole.

Systolic (maximum) blood pressure in the normal adult is in the range of 95 to 140 mm Hg, with 120 mm Hg being the average. These figures are

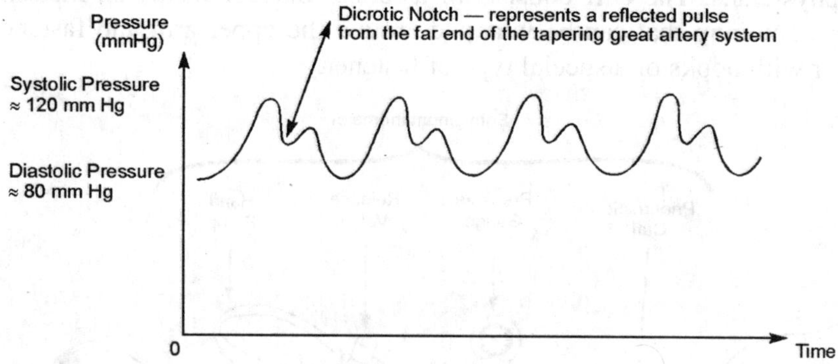

Figure 3.4 *Arterial Blood Pressure Waveform*

subject to much variation with age, climate, dietary habits, and other factors. Normal diastolic blood pressures (lowest pressure between beats) range from 60 to 90 mm Hg with 120 mm Hg being the average. This pressure is usually measured in the brachial artery in the arm. For comparison purposes, with pressures of 130/75 in the aorta—130/75 can be expected in the left ventricle, 9/5 in the left atrium, 25/0 in the right ventricle, 3/0 in the right atrium, and 25/12 in the pulmonary artery. These readings are given as systolic/diastolic pressure.

The mean arterial pressure over one cardiac cycle is approximated by adding one-third of the pulse pressure (difference between systolic and diastolic values) to the diastolic pressure. (All blood pressure measurements are made with reference to atmospheric pressure.) In other words, if P_s is the systolic pressure, and P_d is the diastolic pressure, the mean arterial pressure $\overline{P}$ is given by

$$\overline{P} = P_d + \frac{P_s - P_d}{3}$$

For example, if P_s = 120 mm Hg and P_d = 80 mm Hg, $\overline{P}$ works out to be ≈ 93 mm Hg.

3.3 Measurement of Absolute Arterial Blood Pressure (Korotkoff method)

By far the most common, indirect method of measuring blood pressure involves the use of a sphygmomanometer and a stethoscope. The sphygmomanometer, as shown in Figure 3.5, consists of an inflatable pressure cuff, a hand pump (a rubber bulb) with a release valve and a mercury manometer or pressure gauge to measure the pressure in the cuff, used by all physicians. The cuff consists of a rubber bladder inside an inelastic fabric covering that can be wrapped around the upper arm and fastened either with hooks or a special type of fastener.

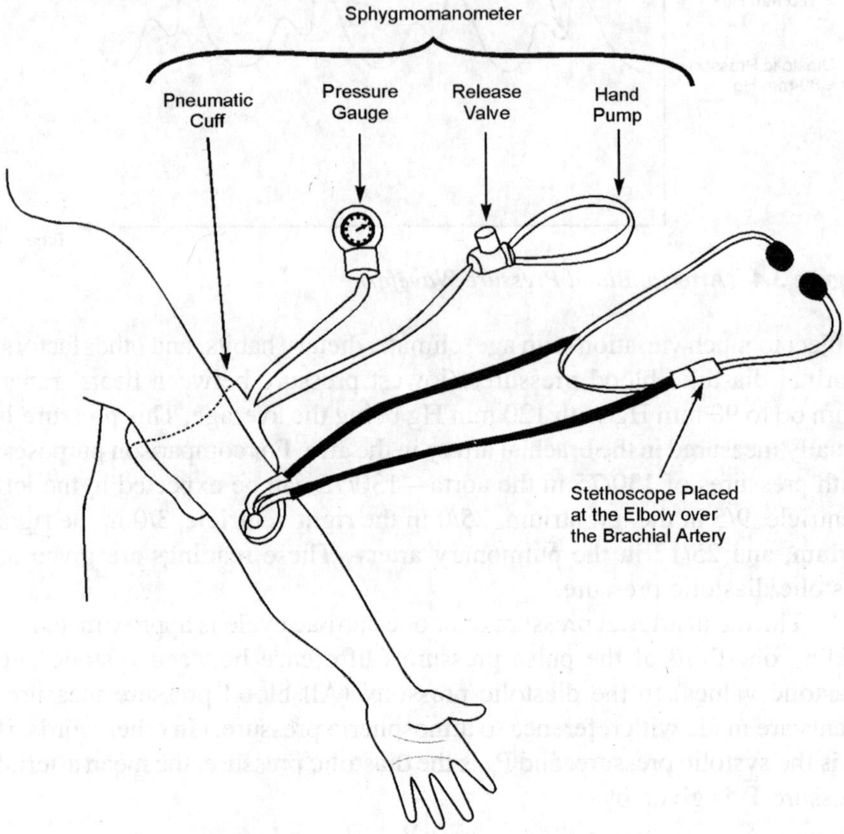

Figure 3.5 *Indirect Blood Pressure Measurement with a Sphygmomanometer*

The sphygmomanometer works on the principle that when the cuff is placed on the upper arm (because a large artery passes down the arm at that point) and inflated, arterial blood can flow past the cuff only when the arterial pressure exceeds the pressure in the cuff. Furthermore, when the cuff is inflated to a pressure that only partially occludes the brachial artery, turbulence is generated in the blood as it spurts through the tiny arterial opening during each systole. The sounds generated by this turbulence, the 'Korotkoff sounds', can be heard through a stethoscope placed over the artery downstream from the cuff.

To obtain a blood pressure measurement, the pressure cuff on the upper arm is first inflated to a pressure well above the systolic pressure, when the artery can be compressed so that the blood flow stops and no sounds can be heard through the stethoscope at this point. The stethoscope is then placed just below the cuff (*i.e.*, "downstream") and the pressure in the cuff is slowly released with the help of the release valve. As soon as the cuff pressure falls below the systolic pressure, small amounts of blood spurt past the cuff and the Korotkoff sounds begin to be heard through the stethoscope. The pressure of the cuff that is indicated on the manometer or pressure gauge when the first Korotkoff sound is heard, is recorded as the "systolic blood pressure". In a healthy adult who is resting, it is normally about 120 mm Hg.

As the pressure in the cuff continues to drop, the Korotkoff sounds continue until the cuff pressure is no longer sufficient to occlude the vessel during any part of the cycle. Below this pressure the Korotkoff sounds disappear, marking the value of the "diastolic pressure" which is normally about 80 mm Hg in a resting adult. The Korotkoff method is also called auscultatory method of sphygmomanometry.

The advantages of this method are that it is simple to perform, involves very little discomfort and is non-traumatic for the patient and repeatable results are obtained. The disadvantages are that the accuracy is not better than 10 mm Hg and that a continuous recording of pressure variations and indication of the details of the pressure waveform cannot be obtained.

With the help of electronics, the standard sphygmomanometer methods have been modified to make the technique less dependent upon the skill of the operator and, in some versions, to enable the subject to take measurements of his own blood pressure. Some of the developments in this direction are as follows.

3.3.1 Oscillometric Method

The key issue in semi-automated and automated blood pressure measurement is the non-auscultatory identification of the systolic and diastolic events, because identifying these events by auscultation requires skill. The oscillometric method of identification is based upon the fact that the volumetric change in blood content in the arm during the cardiac cycle, generates a pressure fluctuation in the blood-pressure cuff which is wrapped round the arm and inflated with air. These pressure variations are small fluctuations which summate with the static cuff pressure imposed by external air pumping. When the cuff is inflated to a pressure above the systolic pressure, the fluctuations in the cuff pressure are small. A microphone placed over the cubital fossa, that is at the bend of the arm, just below the lower margin of the blood-pressure cuff, does not pick up any sound because the cuff pressure occludes the underlying artery and there is no blood flow.

As the cuff pressure is gradually lowered, a point is reached at which the cuff pressure falls slightly below the arterial pressure. Blood begins to flow through the artery in spurts and the turbulence of the blood flow generates a sound which is picked up by the microphone. The microphone is connected to an audio amplifier and the amplified audio signal is rectified and averaged to give a signal which indicates that the cuff pressure has fallen just below the systolic blood pressure. A pressure transducer connected to the cuff then gives a read-out of the systolic pressure.

As the cuff pressure is further lowered, the amplitude of the pulsations in the cuff pressure, on account of the volumetric change in blood content in the arm, progressively increases. These pulsations are in synchrony with the heartbeat. That is, if the heart is beating at 70/min, the pressure fluctuations in the cuff will be at a frequency of 70/min, and these variations will summate with the steady air pressure inside the cuff. A complex phenomena of pressure–volume interactions prevail. Studies have shown that with the further lowering of the cuff pressure, the fluctuations reach the maximum when the cuff pressure is the "mean pressure" which is [(systolic pressure + 2 × diastolic pressure)/3]. The factor '2' with the diastolic pressure arises on account of the fact that the diastolic time period is typically twice as long as the systolic time period.

The cuff pressure, as given by the pressure transducer, is sampled at intervals of about 1 ms and by means of an on-line microprocessor each sample value is compared to the immediate past sample value. By this means, the peak of each of the fluctuation cycles and the overall maximum of the fluctuations is identified. The automated system records the cuff pressure, when the 'mean pressure' is reached, as depicted by the above criteria. Thus, the systolic pressure and the mean pressure are determined, and, by the relationship between the pressures given earlier, the diastolic pressure is automatically computed by the microprocessor system. In the oscillometric method, therefore, the diastolic event is not as such identified and the diastolic pressure is determined by an 'estimation' technique.

The above description gives a 'semi-automated' technique for blood-pressure measurement. The cuff inflation and controlled lowering of the cuff pressure is manually done. All the other activities are automated and do not require any special operator skill. More advanced versions of blood pressure measuring instruments have automated air pressure pumps and servo-controlled bleed valves, by means of which all the manual processes are eliminated. Such systems are commonly in use in intensive coronary care wards to monitor seriously ill cardiac patients. Pressure measurements are performed automatically at specified intervals and whenever the pressure is beyond safe limits, an alarm alerts the nursing staff.

3.3.2 Doppler Technique

Another approach to the automated recognition of systolic and diastolic events is by means of the Doppler technique. A Doppler probe, which is essentially a piezoelectric crystal which emits a vibration at a certain frequency and receives the reflected vibrations, is incorporated in the blood-pressure cuff. The blood flow velocity data, as determined by methods described in a later chapter, is subjected to 'pattern analysis' by an online microprocessor system. There are specific characteristics of the flow at the time when the cuff pressure is just below the systolic pressure and when the cuff pressure is at the diastolic level. The events are thus identified by the pattern recognition operation and the cuff pressures at those states are displayed.

3.3.3 Finger Blood Pressure Gauge

Instead of placing the pressure cuff on the upper arm, the pressure cuff can also be placed on the finger. The configuration is so made that the finger, when slipped into a small cavity, gets surrounded by a cuff. There are a number of ways to automatically recognize the systolic and diastolic events. One approach is based upon 'photoplethysmography'. A narrow light beam, obliquely incident on the surface of the finger opposite to the nail, is partly absorbed and partly reflected. The reflectance is a function of the total blood content of the finger. Shortly after the systolic event, the blood volume in the finger reaches a maximum and the volume a minimum in relation to the diastolic event. Hence, by sensing the light reflectance by means of a phototransistor and analyzing the magnitudes, the systolic and diastolic events can be identified and correlated with the cuff pressure to give an automated indication of the pressures as done in the preceding techniques.

3.4 Electrical Potentials Generated within the Heart—Origin of the Electrocardiogram Waveform

The preceding section covers the mechanical activity of the heart and states that mechanical activity is initiated by contraction of the muscles surrounding the atria. In other words, it may be stated that the muscle contraction is initiated by stimulation.

The right atrium of the heart contains a bundle of nerves known as the Sinoatrial node (SA node). This type of nerve cell is not found anywhere else in the body. Its function is to start the heartbeat and set its rhythm or pace; this node is also called a "cardiac pacemaker" and generates impulses at the normal rate of the heart, about 70 beats per minute at rest. The electrical and mechanical output from the heart is initiated by stimulation from this node which results in contraction of the various heart muscles.

Impulses generated by the SA node stimulate contraction of the muscles comprising the atria. These impulses also travel along conducting fibres in the atrium to the Atrioventricular node or AV node, stimulating the depolarization of this node. The AV node is located in the lower part of the heart-wall between the two atria on the septum, and acts as a *delay line* to provide timing between the action of the atria and ventricles. Stimulation of the AV node causes impulses to be sent to the myocardium or muscles comprising the ventricles via the Bundle of His, two bundle branches on each of the septums and the fine Purkinje fibres, which arborize in the ventricular muscle. Thus, the atria and ventricles are functionally linked only by the AV node and the conduction system. The AV delay is provided so that the atrial contraction can complete the ventricular filling before the contraction of the ventricles. So, the muscular contractions necessary to maintain the heart's pumping action are initiated by depolarization and repolarization of the SA node and then depolarization and subsequent repolarization of the AV node. Also, when the ventricles are depolarizing, the atria are repolarizing. These depolarizations and repolarizations generate external action potentials which can be recorded at the surface of the body. These external action potentials generated from within the heart are known as the electrocardiogram or ECG. The ECG waveform (Figure 3.6) is shown in relationship to the mechanical action of the heart and the resultant arterial pressure.

3.5 Composition of the ECG

The electrical activity of the heart is, as stated earlier, initiated by depolarization of the SA node and a resulting contraction of the muscles surrounding the atria. The resulting external action potential is known as the P-wave. Immediately following this depolarization, repolarization of the atria occurs; however, for some reason, this does not generate a pronounced action potential. This potential is known as the TA-wave and

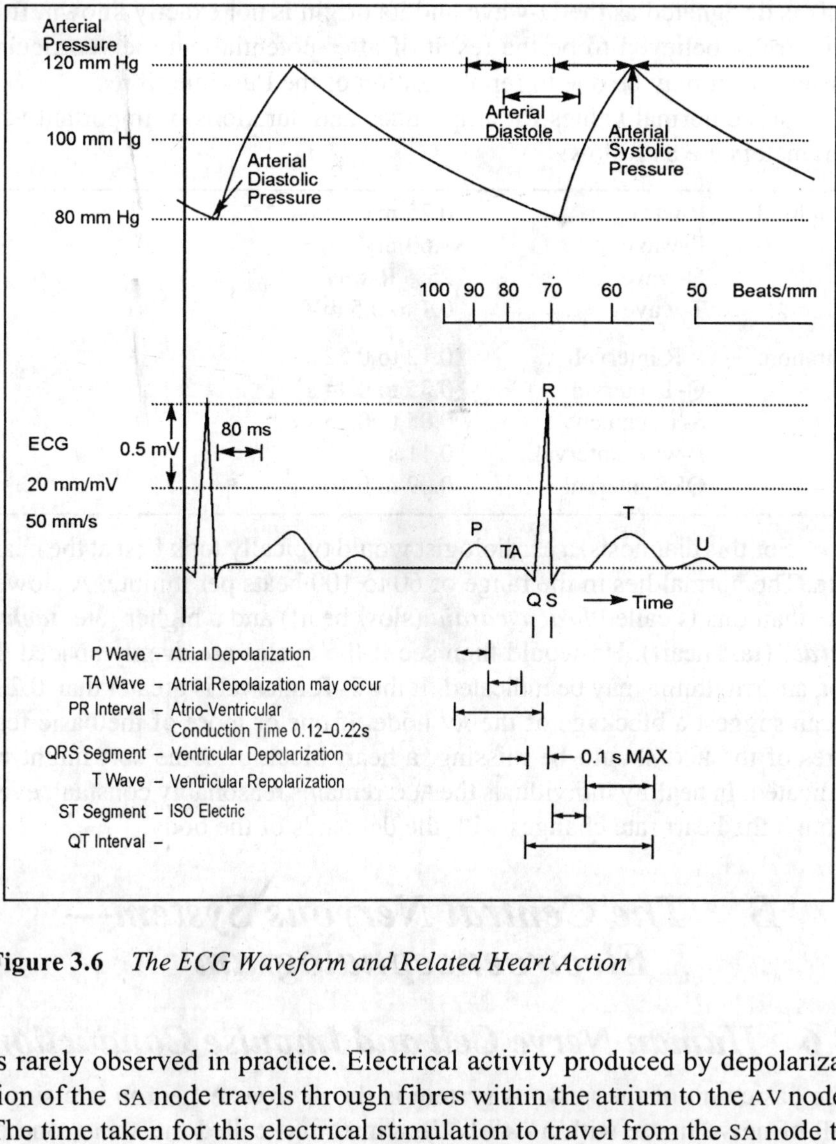

Figure 3.6 *The ECG Waveform and Related Heart Action*

is rarely observed in practice. Electrical activity produced by depolarization of the SA node travels through fibres within the atrium to the AV node. The time taken for this electrical stimulation to travel from the SA node to the AV node is known as the atrioventricular conduction time and is typically between 120 ms and 220 ms. When this stimulation reaches the AV node, this node depolarizes and the depolarization is conducted down through the bundle of His to the myocardium muscle causing ventricular depolarization. The external action potential is referred to as the QRS complex. Immediately following this depolarization, the cells concerned repolarize which results in ventricular repolarization or the T-wave. Many ECG waveforms also show an additional wave occurring after the T-wave.

This is designated as the U-wave and its origin is not exactly known. It is gene- rally believed to be the result of after-potentials in the ventricular muscles or it may be due to repolarization of the Purkinje fibres.

Some normal values for amplitudes and durations of important ECG parameters are as follows:

Amplitude:	P-wave	0.25 mV
	R-wave	1.60 mV
	Q-wave	25% R-wave
	T-wave	0.1 to 0.5 mV
Duration:	P-R interval	0.12 to 0.22 s
	Q-T interval	0.35 to 0.44 s
	S-T segment	0.05 to 0.15 s
	P-wave interval	0.11 s
	QRS interval	0.09 to 0.10 s

For this diagnosis, a cardiologist would typically look first at the heart rate. The normal lies in the range of 60 to 100 beats per minute. A slower rate than this is called *brachycardia* (slow heart) and a higher rate, *tachycardia* (fast heart). He would then see if the cycles are evenly spaced. If not, an arrhythmia may be indicated. If the P–R interval is greater than 0.2 s, it can suggest a blockage of the AV node. If one or more of the basic features of the ECG should be missing, a heart block of some sort might be indicated. In healthy individuals the ECG remains reasonably constant, even though the heart rate changes with the demands of the body.

B. The Central Nervous System—Electroencephalogram

3.6 Human Nerve Cell and Impulse Conduction

The basic unit of the nervous system is the *neuron*. A neuron is a single cell of nerve tissue with a *cell body*, sometimes called the *soma*, one or more "input" fibres called *dendrites*, and a long transmitting fibre called the *axon*. Often, the axon branches near its end into two or more terminals, as shown in Figure 3.7.

The portion of the axon immediately adjacent to the cell body is called the axon hillock. This is the point at which action potentials are usually generated. Certain types of neurons have axons or dendrites coated with a fatty insulating substance called myelin. The coating is called a *myelin sheath* and the fiber is said to be *myelinated*. In some cases, the myelin

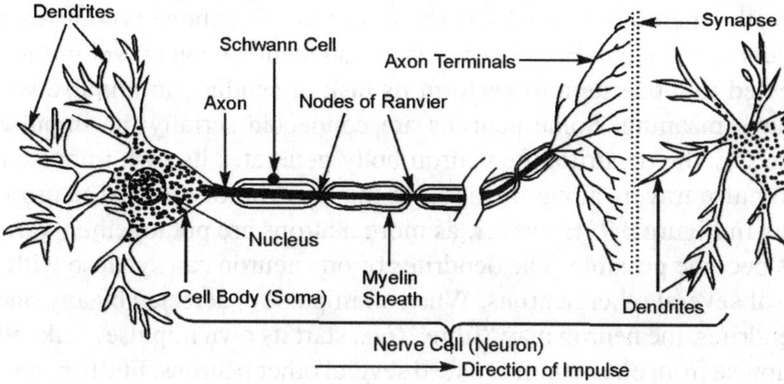

Figure 3.7 *Human Nerve Cell and Impulses*

sheath is interrupted at rather regular intervals by the *nodes of Ranvier*, which help speed up the transmission of information along the nerves. Outside of the central nervous system, the myelin sheath is surrounded by another insulating layer, called the *neurilemma*. This layer, thinner than the myelin sheath and continuous over the nodes of Ranvier, is made up of thin cells, called *Schwann cells*.

The dendrites, the thread-like structure of a neuron, are very sensitive to stimulus. The neurons can receive stimuli and pass it along (technically *conduct*) in the form of an electrochemical impulse. When a stimulus is detected by a dendrite, an impulse is generated in the soma, and sent along the axon attached to the membrane. In this way, impulses are transmitted through nerve tissue. Neurons are *not* connected directly to each other.

Impulses are transmitted from one neuron to another via a "synapse" (cell interconnection). It is at this point that the action potential mechanism is utilized from the axon of the neuron, across the synapse, to a dendrite of another neuron. There are two types of neurons: *sensory*, which receive a stimulus and transmit it to a central control station for analysis, and *motor*, which transmit an impulse to a muscle tissue for action to be possible.

3.7 The Nervous System

The nervous system is at once the most complex, the most simple, and the least understood system in the body. Only a rudimentary discussion will be attempted here.

The basic operating principle of the nervous system is the action potential as discussed previously in the section on cells. An action potential can be considered as an impulse. Each nerve cell or neuron is connected to all the other neurons in much the same way as a telephone is connected to the telephone system. Each neuron is an independent agent; yet it must be connected to the system to perform its task of sending an impulse which will have meaning. Some neurons are connected serially; an impulse is detected by the dendrite, the neuron body generates its own impulse and that impulse travels along the axon to the dendrite of the next neuron, as we saw in Figure 3.7. However, as more neurons are put together, various routes become possible. The dendrites of one neuron can synapse with the axons of several other neurons. When an impulse is detected by any one of the dendrites, the neuron may "fibre" (*i.e.*, start its own impulse). Likewise, the impulse from one neuron can feed several other neurons. Further, several dendrites of one neuron may be connected to several axons which individually may not be able to fire the neuron, but collectively can fire it.

Like a telephone system or computer, there are two sides to the nervous system; the input and the output. Neurons, however, can perform both functions. For input, the sensory neurons send an impulse to a "central station" such as the spinal cord; if analysis is required, the signal will go up to the "central analyzer", the brain. Output takes the form of another impulse going out of the brain or spinal cord via a motor neuron to a muscle or another organ.

The main function of the nervous system is the control of bodily activity through the monitoring of all the previously discussed systems and their status. When some parameter changes, an impulse is generated by a sensory neuron. Some of these impulses have a higher priority than others, resulting in various levels of response, from immediate to none. For some stimuli, a person has "conscious" control over his responses; others are self-responsive (*e.g.*, in the former: body movements; in the latter: the heart rate, breathing, digestion).

There are also some checks and balances in the nervous system. Some organs have both motor neurons which are designed to increase activity and motor neurons which are designed to decrease activity. It is the body's needs at any particular moment which will determine which motor neurons have more effect (*e.g.*, the iris of the eye being larger or smaller, the heart beating slower or faster).

The spinal cord serves to channel signals for *reflexive actions*, *e.g.*, with heat applied on a finger, the contraction of hand muscles; urinating; defecating and the response of the body to environmental temperature changes.

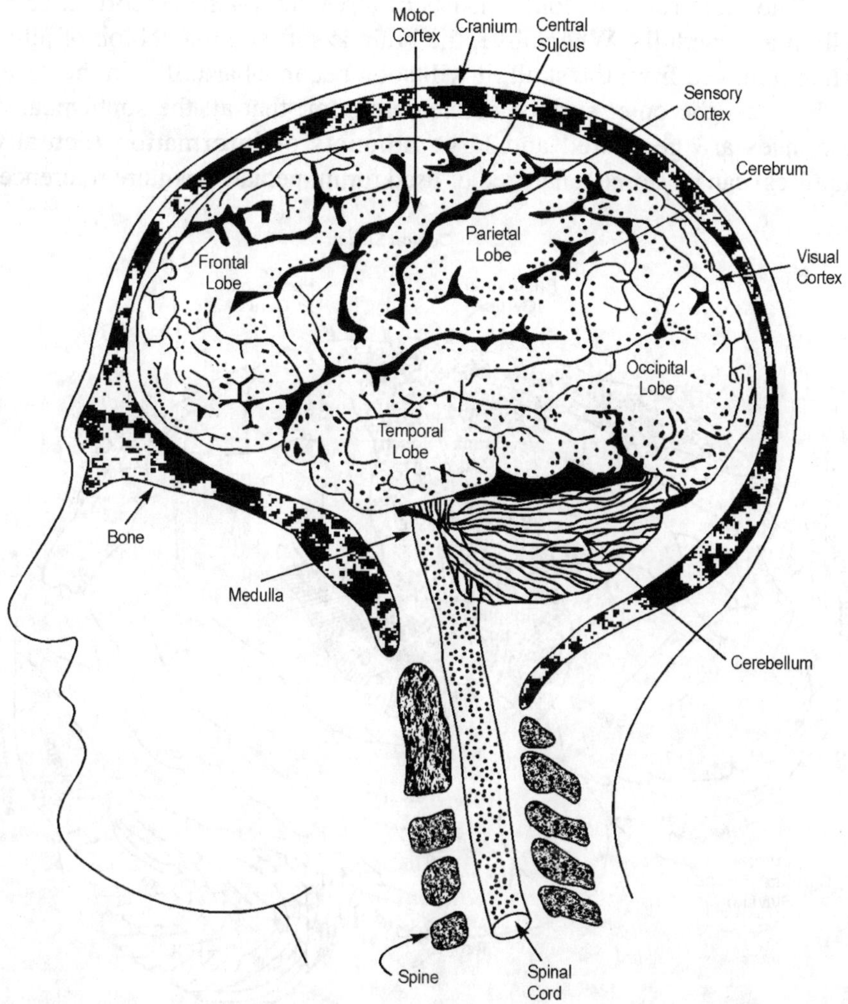

Figure 3.8 *Anatomical Sections of the Brain*

3.8 Anatomy of the Brain

A very general look at the anatomy of the brain should be helpful in the understanding of the functions of the central nervous system. Figure 3.8

shows the main parts of the brain and spinal cord and Figure 3.9 gives a broad view of the cerebral cortex.

The brain (encephalon) consists of three parts—the cerebrum, cerebellum and medulla. When alive, the brain is soft and like a blob of jelly. When removed from the skull, it will soon become hard. It is in the *cerebral cortex* (the outer surface of the cerebrum) that all the sophisticated responses are generated and large amounts of information (actually impulses) can be stored, analyzed or used for immediate or future reference.

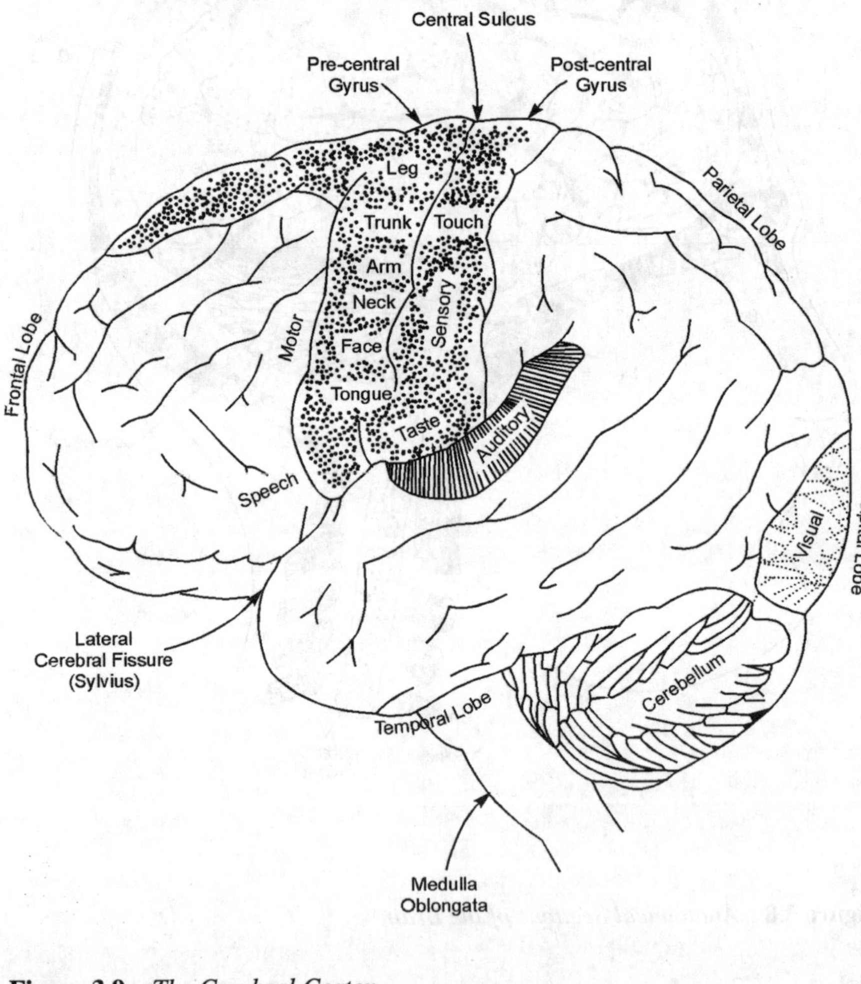

Figure 3.9 *The Cerebral Cortex*

The cerebral cortex contains some 9 billion of the 12 billion neurons found in the human brain. It has been experimentally determined that in this section are located: (1) various sensing activities (seeing and hearing, in a sense, understanding what is seen and heard), (2) the sense of feeling in various parts of the body from head to toes, (3) the ability to recall past events, (4) the sophisticated emotions, and (5) the thinking process. The cerebellum acts as a physiological microcomputer which intercepts various sensory and motor nerves to smooth out what would otherwise be "jerky" muscle motions. It also plays a vital role in man's ability to maintain his balance. The medulla seems to be associated with the control of some of the basic functions responsible for life, such as breathing, heart rate, and kidney functions. For this purpose, the medulla contains a number of timing mechanisms, as well as important connections, so it is believed.

The cerebrum consists of two hemispheres, separated by a deep fissure called the *corpus callosum*. The hemispheres are divided into four lobes: the frontal lobe, parietal lobe, occipital lobe and temporal lobe. Briefly, the frontal lobe is for intelligence. The upper side of the temporal lobe consists of the hearing centre. In the posterior part of the occipital lobe, the vision centre is situated. In the anterior part of the parietal lobe, there are the sensory centre and the motor centre. The temporal lobes are for the storage process in long-term memory.

3.9 Excitation and Inhibition Potentials

In Chapter 2, cells were referred to as bistable devices existing in only two states; a polarized state of –90 mV and a depolarized state of +20 mV. Intermediate potentials were regarded as having little significance. The nerve cells in the central nervous system and the brain exist in a polarized state between –70 mV and –110 mV, and the exact cell potential within this range appears to be significant as it determines the cell's vulnerability to a stimulus that would result in a regenerative breakdown.

Consider a group of cells producing zero activity; all cells within this group would rest at the normal polarized level of –90 mV. Assuming that one cell is depolarized from an external stimulus, this depolarization will affect the adjoining cells via the synapses, causing a change in the resting potential of these adjoining cells. The effect on these adjoining cells can be either to raise or to lower their resting potential, depending on the type of fibre involved in the synaptic junction. If the resting potential is increased, a breakdown of the cell membrane will not occur and the impulse can be thought of as having had an *inhibitory effect*. If the impulse arriving at this synapse decreases the resting potential, then the cell is more susceptible to depolarization and the impulse can be thought of as having had an *excitatory effect*. With repeated excitatory stimuli at a sufficiently high rate, the membrane will eventually fully depolarize and a new impulse will be propagated. Any single cell is influenced by synapses from many other cells; some are excitatory and some inhibitory. The play of these two opposite effects on any area may or may not cause depolarization of the cell; the deciding factor is the number of impulses being received per unit time and the balance between excitation and inhibition; thus, frequency becomes an important parameter. Figure 3.10 shows the effect of various excitation and inhibition signals of any one individual cell, finally resulting in depolarization of the cell once the –70 mV threshold is obtained.

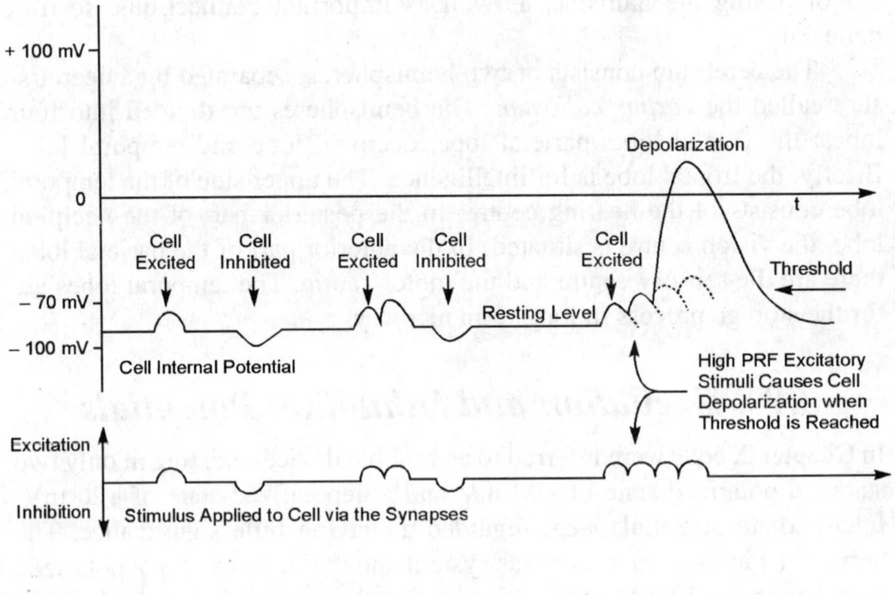

Figure 3.10 *Stimulus Changing Cell Resting Potential may Result in Depolarization of the Cell*

The effect of the various neurons acting on a receiving neuron are reflected in changes in the graded potentials of the receiving neuron. *Graded potentials* are variations around the average value of the resting potential. When this graded potential reaches a certain threshold, the neuron fires and an action develops. Regardless of the graded potential before firing, the action potentials of a given neuron are always the same and always travel at the same rate. An excitatory graded potential is called an *excitatory post-synaptic potential* (EPSP), and an inhibitory graded potential is called an *inhibitory post-synaptic potential* (IPSP). Synapses behave much like multiple-input AND and NOR logic gates and, by their widely varied patterns of excitatory and inhibitory "connections", provide a means of switching and interconnecting parts of the nervous system with a complexity far greater than anything yet conceived by man.

3.10 The Electroencephalogram (EEG)

At this stage, the brain can be regarded as a highly developed biochemical factory. The electrical activity which results from so much chemical change is known as the electroencephalogram (EEG) and is, in a sense, a useful by-product of nervous action since it allows us to make non-mutilating measurements on an organ which is singularly resentful of external interference. As a result of the neurons of the brain "firing", electrical signals or biopotentials are generated. With sensitive monitoring equipment, these potentials can be detected and analyzed. This is accomplished by placing electrodes at strategic points around the head, and making a recording of these firings. The EEG, as recorded from the surface of the head, consists of rhythmical, slow, sinusoidal-like waveforms between 10 and 100 microvolts in amplitude. The EEG varies in form, amplitude and frequency; the basic frequency of around 10 Hz is known as the alpha rhythm.

It is not the firing of individual neurons that is recorded in an EEG. Rather it is the "gross" electrical activity that is monitored. Because the recordings appear as many different waveforms, they are commonly referred to as "brain waves". These waveforms are then analyzed, mainly on the basis of many other persons' recordings.

Brain waves themselves are categorized by frequency, because of the meaning that is revealed at these different frequencies. Five main divisions (Frequency Bands) have been established.

Frequency Band	*Range*
Delta	0.5 – 4.0 Hz
Theta	4.0 – 8.0 Hz
Alpha	8.0 – 13.0 Hz
Beta	13.0 – 22.0 Hz
Gamma	22.0 – 30.0 Hz

The most important wave is the alpha because of its relation to mental activity. Normally, with an increase in mental activity, the alpha will reduce markedly. However, it is not uncommon to find this frequency missing in children and some healthy normal adults. The EEG produced by a subject in various conditions is shown in Figure 3.11.

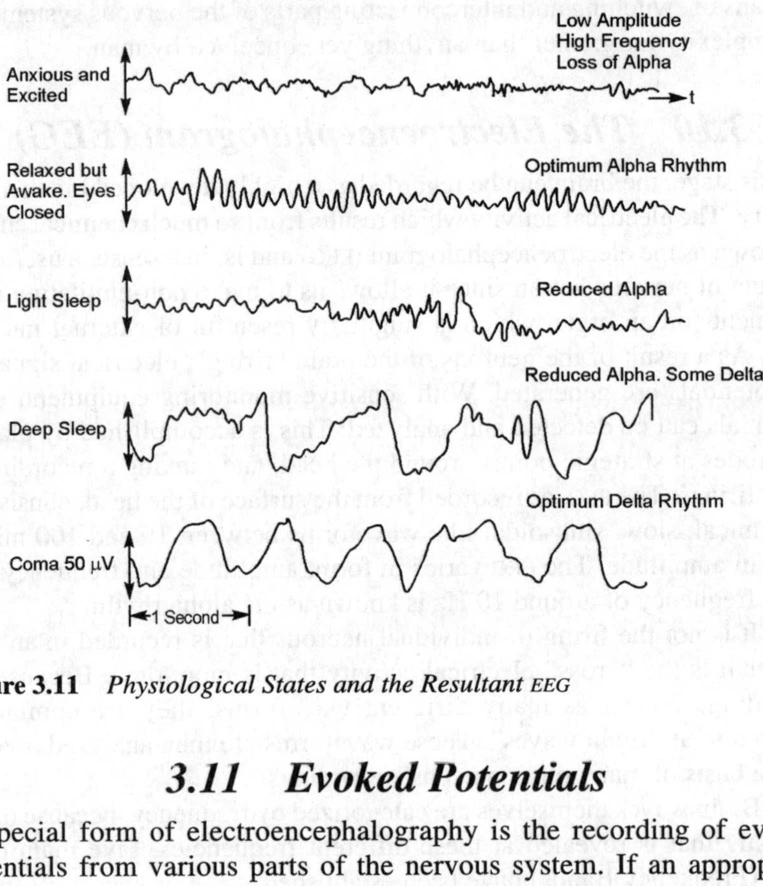

Figure 3.11 *Physiological States and the Resultant* EEG

3.11 Evoked Potentials

A special form of electroencephalography is the recording of evoked potentials from various parts of the nervous system. If an appropriate stimulus is applied to a sense receptor, the corresponding sensory area of the brain responds by producing an electric potential known as the *evoked potential*. The evoked potential, as it appears at the surface of the brain, is the integrative result from the action of many cells. When recorded externally on the scalp, the evoked potential is of the order of 10 µV or so. It is not yet completely known which cortical elements are responsible for these potentials. Often, they are camouflaged by the EEG and, thus, it may be necessary to remove the EEG by an averaging technique when attempting to record the evoked potentials.

Nowadays, the term "event related potential" has been used instead of evoked potential. This is because there are some changes that are evoked by an external stimulus but are related to an event. The studies on evoked potential are more prone to contamination by artifact than the clinical EEG. The artefact is synchronized with the stimulus and eventually appears with the evoked potential. Therefore, good recording techniques are to be used for evoked potential studies.

C. Muscle Action and the Sensory System—Electromyogram (EMG)

Fundamental relationships exist between muscle function, the sense receptors, the brain, the nervous system and the peripheral nerves. The discussion of muscle function that follows includes a discussion of the physiology of the nervous system and the peripheral nerves. The brain is, at this stage, regarded as a processor.

3.12 The Motor Unit

The motor unit, as the name implies, is the biological unit of muscle function. A motor unit consists of a motor nerve arising from motor neurons in the brain-stem or spinal cord and branching into various motor end-plates. These motor end-plates are each connected to an individual muscle fibre; stimulation of these motor end-plates causes the contraction of the single muscle fibre attached to it, as shown in Figure 3.12. The number of motor units varies between the different muscles of the body; generally speaking, the larger the muscle, the more motor units will be found in that muscle. The size of the motor unit, that is, the number of muscle fibres activated by

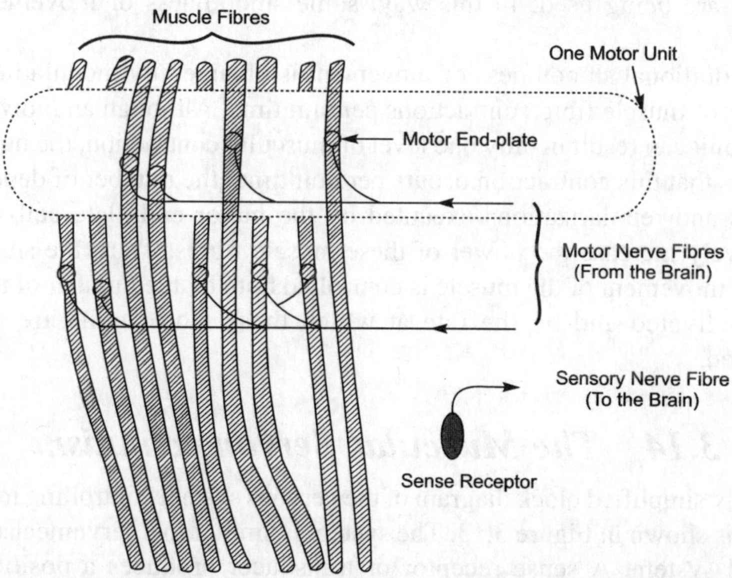

Figure 3.12 *Relationship Between Nerve and Muscle*

the same nerve fibre, can be quite different for different muscles; in man, one motor unit may contain from 25 to 2 000 muscle fibres. The force developed by a motor unit may range from 0.1 to 250 gm in weight. The muscle fibres of a motor unit are not clumped together in one part of the muscle, but rather, the muscle fibres of different units are interlaced as shown in Figure 3.12.

3.13 Muscle Action

Chapter 2 stated that a cell can exist in its polarized or depolarized state; that is, it is a bistable device and its intermediate potential levels are not stable. Motor nerves are also cellular in nature. Thus, any individual nerve can only exist in polarized or depolarized state and will transmit only two potential levels to the motor end-plates causing a bistable "on-off" action of the muscle fibres. So, the individual muscle fibres of one motor unit can exist in only two states, a relaxed state or a tensed state. Normal muscular activity is characterized by the smoothness of movement, steadiness and precision. These characteristics are due to the large number of motor units comprising any one muscle. If a small muscular effort is required, only one motor unit will be called into action; as increasing muscular effort is required, many more motor units are called into action until the muscle is provided maximum effort, at which time all motor units connected to this muscle are being used. In this way, some smoothness of movement is obtained.

Additional smoothness of movement is obtained by modulating the number of muscle fibre contractions per unit time. Although an individual motor unit can result in only one level of muscular contraction, the number of times that this contraction occurs per unit time (the number of depolarizations and repolarizations executed by the motor end-plate cells) will effectively increase the power of these muscle fibres. Thus, the smoothness of movement of the muscle is controlled both by the number of motor units activated and by the rate at which these motor units are being activated.

3.14 The Muscular Servomechanism

A highly simplified block diagram of the nervous system controlling muscle action is shown in Figure 3.13. The system is similar to a servomechanism control system. A sense receptor or transducer produces a position or velocity signal, which is sent to the brain via the sensory nerve. The brain, in turn, initiates an "error" or control signal by comparing the measured position with the desired one stored in the memory. This signal is sent via the motor nerve to the muscle to control its action. An emergency "gate" has been shown in Figure 3.13, this emergency gate is not markedly

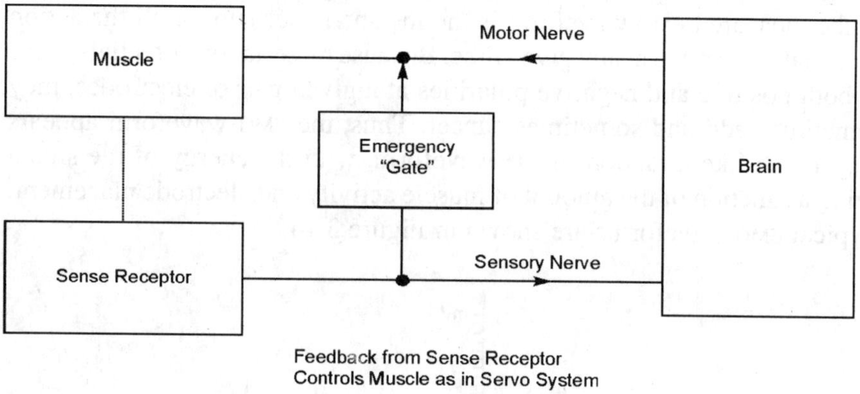

Figure 3.13 *Block Diagram of the Nervous System*

involved in normal receptor/muscle operation. When a reflex response is called for, however, this emergency gate bypasses the signal path to and from the brain and initiates a reflex response. This emergency "gate" is usually located within the spinal cord. Thus, a reflex response results from a "large" signal, *i.e.*, a high repetition rate signal being received from a sense receptor. This signal bypasses the brain in the initiation of muscle action. These reflex responses protect the body from serious damage.

3.15 The Potential Generated During Muscle Action—Electromyogram (EMG)

Depolarization initiated within a sense receptor travels to the brain along the sensory nerve fibre as a series of travelling depolarization waves. The brain then initiates another series of travelling waves of depolarization along the motor nerves to cause a series of depolarizations of the motor end-plates. Depolarization of the motor end-plates depolarizes the cells within the muscle fibre, causing contraction of these fibres. The actual internal cell potentials involved here are the normal cell polarized potential of –90 mV and the normal cell depolarized potential of +20 mV.

The bioelectric potentials associated with muscle activity constitute the *electromyogram* (EMG). These potentials may be measured at the surface of the body near a muscle of interest or directly from the muscle by penetrating the skin with needle electrodes. Since most EMG measurements are intended to obtain an indication of the amount of activity of a given muscle or group of muscles, rather than of an individual muscle fibre, the pattern is usually a summation of the individual action potentials from the fibres constituting the muscle or muscles being measured. The amplitude of the measured EMG waveform is the instantaneous sum of all the action potentials generated at any given time. Because these action potentials occur in both positive and negative polarities at a given pair of electrodes, they sometimes add and sometimes cancel. Thus, the EMG waveform appears very much like a random-noise waveform, with the energy of the signal being a function of the amount of muscle activity and electrode placement. Typical EMG waveforms are shown in Figure 3.14.

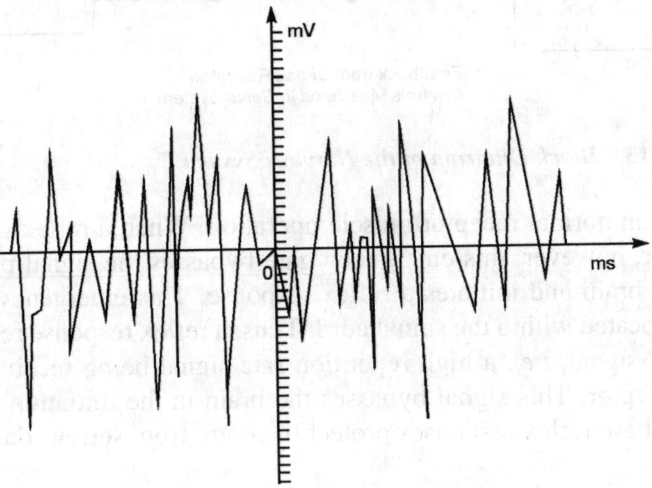

Figure 3.14 *Typical Electromyogram Waveform. EMG of Normal "Interference Pattern" with Full Strength Muscle Contraction. Sweep Speed is 10 ms/cm. Amplitude is 1mV/cm*

Chapter 4

Biopotential Electrodes

Bioelectric events, before they can be put into the amplifier for subsequent record or display, have to be picked up from the surface of the body. This is done by using electrodes. Electrodes play an important role in making satisfactory records of bioelectric signals and their choice requires careful consideration. They should be comfortable for the patients to wear over long periods of time and should not produce any artefacts. Another desirable factor is the convenience of application of the electrodes.

The function of biopotential electrodes is to provide an interface between the body and the electronic instrumentation system. These electrodes, in a way, function as transducers; since the flow of current in the body is through ions, while it is in the form of electrons in the electrode and its lead wire. Thus, the transduction performed by the electrodes is from an ionic current into an electrical current. This introduces the electrode–electrolyte interface process, electrode characteristics and different forms of electrodes to be used for measuring ECG, EEG, EMG and other biopotentials.

4.1 The Electrode–Electrolyte Interface

When electrodes in their simplest form made of pieces of metal, are placed on or inside the body, they come in contact with body fluids which may be considered as electrolytes. Due to this contact between a metal and an electrolyte solution, an electrochemical reaction produces a difference of potential between the metal and solution. The chemical reactions that occur between metals and electrolytes influence the performance of biopotential electrodes.

Figure 4.1(a) shows a silver electrode in an electrolyte (salt solution). Since silver, a good conductor of electricity, has an abundance of loosely held free valence electrons, some of these enter the solution, making the neutral electrode positively charged with respect to the electrolyte. This results in a potential difference called a "half-cell potential". The value of this half-cell voltage for a silver electrode is approximately 0.8 V and for a copper electrode it is approximately 0.3 V. Figure 4.1(b) shows both

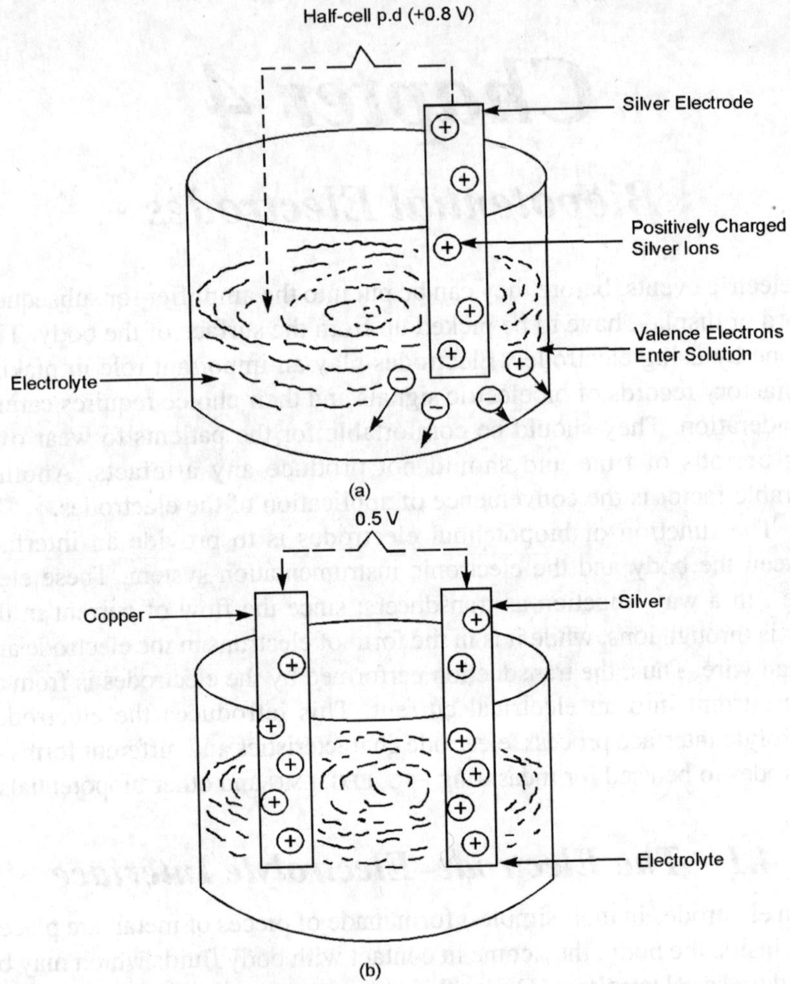

Figure 4.1 *Metal–Electrolyte Interface*
(a) Silver Electrode Half-cell Potential
(b) Terminal Voltage Between Silver and Copper Electrodes Half-cell Potential

the electrodes placed in the electrolyte and, since the half-cell potentials for the silver and copper electrodes are 0.8 V and 0.3 V respectively, both the electrodes being positive with respect to the electrolyte, the potential difference between the electrodes equals 0.5 V. The knowledge of the half-cell potential, which is determined by the metals involved, the concentration of its ions in solution and the temperature are important for understanding the behaviour of biopotential electrodes. Some sort of separation of charges exist at the metal–electrolyte interface which results in an *electrical double layer*, wherein one type of charge is dominant on the surface of the metal and the opposite charge is distributed in excess in the electrolyte lying adjacent to the electrode.

Figure 4.2 shows the cross-section of two silver plates used as biopotential electrodes on the surface of the skin, which act as an electrolyte. The metals will develop equal half-cell potentials; if chemically identical, the resulting potential difference between electrodes will equal zero. The difference of potential between the terminals of biopotential electrodes in contact with the body is called the *electrode offset voltage*. The offset would be zero if the electrodes are chemically identical. However, in practice this is not so and some value of the offset voltage is usually present

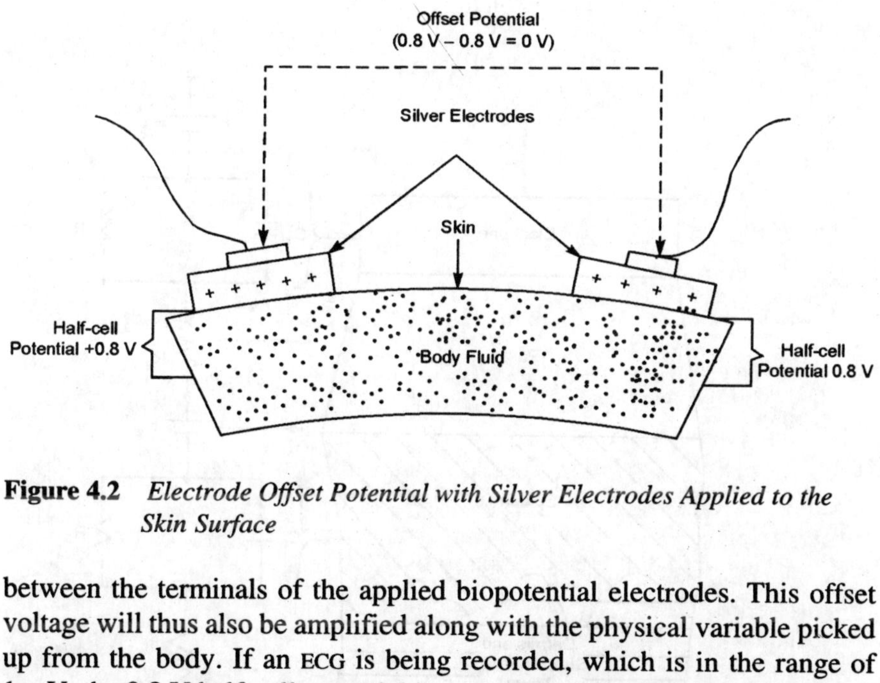

Figure 4.2 *Electrode Offset Potential with Silver Electrodes Applied to the Skin Surface*

between the terminals of the applied biopotential electrodes. This offset voltage will thus also be amplified along with the physical variable picked up from the body. If an ECG is being recorded, which is in the range of 1 mV, the 0.8 V half-cell potential is approximately 1000 times greater. A large offset voltage due to chemically unmatched electrodes, may thus interfere with the desired biosignals and result in undesired artefacts. Both the half-cell potential and noise (voltage fluctuations) from the electrodes, due to chemical activity taking place within them, may be reduced by a proper choice of electrode material and by chloriding the silver electrode—called the *silver-silver-chloride* (*Ag–AgCl*) electrode. The coating of chloride is done chemically by placing two pieces of pure silver (one of them the electrode) in a bromide-free sodium chloride solution (of 5% concentration) and connecting the two pieces to a DC voltage, the positive being connected to the electrode to be chlorided. Chloride ions from the salt solution then combine with the silver ions of the electrode and a thin film of chloride molecules is deposited on the silver electrode making it look grey. Chloriding is best done in a dark room for a duration of about three minutes with a 3 volt battery.

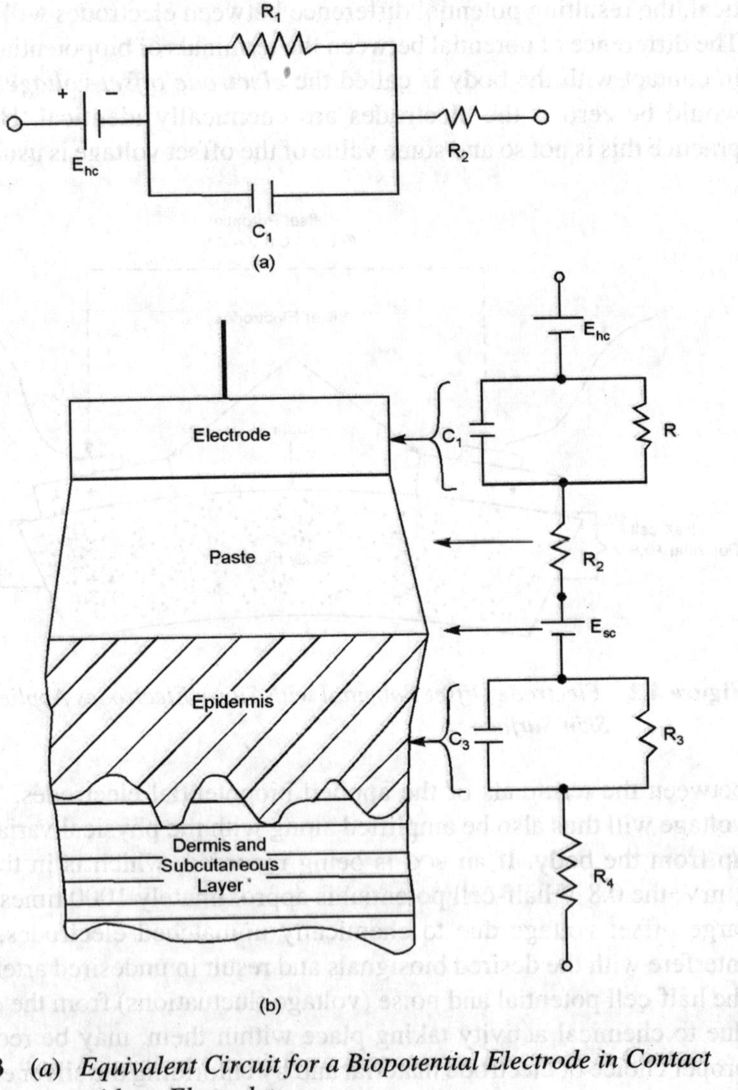

Figure 4.3 *(a) Equivalent Circuit for a Biopotential Electrode in Contact with an Electrolyte*
(b) Total Electrical Equivalent Circuit between the Electrode and the Skin

Some variations in the process are used to produce electrodes with specific characteristics.

When an electrode is placed on the skin surface, there is some electrical resistance at the electrode–skin interface. Since the skin's natural resistance is high compared to the resistance of the body fluids, the selected skin site is to be well prepared by cleaning with alcohol or acetone and by applying a commercially available conductive jelly (electrode paste). This ensures a low value of electrode–skin interface resistance. Surface electrodes usually have resistances of 2 000 to 10 000 ohms depending on their size, whereas, small needle electrodes have a much higher resistance.

A simple series equivalent circuit of an electrode–electrolyte interface is shown in Figure 4.3(a). Here, E_{hc} is the half-cell potential; R_1 and C_1 represent the impedance associated with electrode–electrolyte interface; and R_2 is the total series resistance in the circuit due to resistance in electrolyte and electrode lead wire.

However, when an electrode makes contact with the skin via an electrolyte paste, the equivalent circuit is modified as shown in Figure 4.3(b). Now, R_2 becomes the effective resistance of the paste between the electrode and the skin. The epidermis of the skin may be considered as a semi-permeable membrane and the potential difference across it is represented by E_{se}. The epidermic layer has also an electrical impedance, which is represented by the parallel circuit C_3, R_3. The dermis and subcutaneous layer under it behave in general as pure resistance R_4, as shown in Figure 4.3(b). Thus, it can be seen that to obtain a more stable electrode, the effect of the epidermis (stratum corneum) has to be reduced. This is achieved by many ways: by rubbing with a pad soaked in acetone or by puncturing the epidermis with dental burrs. All these methods tend to short out E_{se}, C_3 and R_3, thus improving the stability of the signal.

4.2 Electrode Types

A wide variety of electrodes can be used to measure bioelectric events, but nearly all can be classified as belonging to one of three basic types:

1. Surface Electrodes: Used to measure ECG, EEG and EMG potentials on the surface of the skin.
2. Needle Electrodes: Used to penetrate the skin to record EEG potentials from a local region of the brain, or EMG potentials from a specific group of muscles.
3. Microelectrodes: Used to measure bioelectric potentials near a single cell.

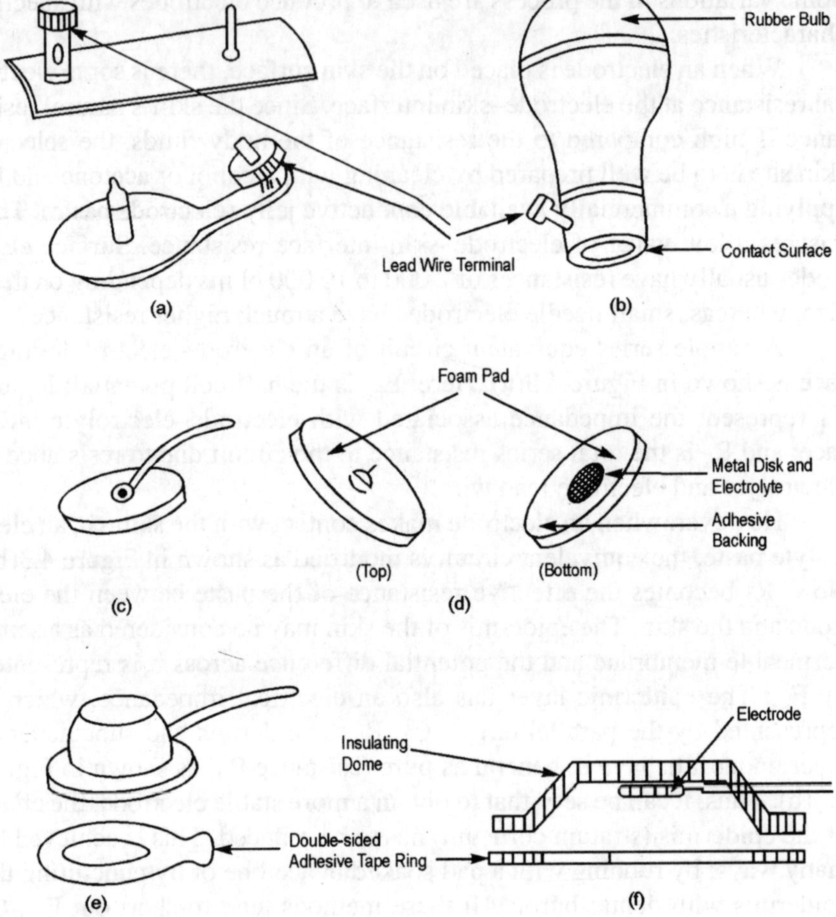

Figure 4.4 *Body Surface Metal Plate Biopotential Electrodes*
(a) Normally used for ECG
(b) Suction Electrode used as a Precordial Electrode for ECG
(c) Chest Electrode Attached by Surgical Tape for Long-term ECG
(d) Disposable Foam Pad Electrode for ECG
(e) & (f) Floating Electrode

4.2.1 Surface Metal Plate Electrodes

One of the most commonly used devices for recording bioelectric events are the metallic plate electrodes. One such type is frequently employed in electrocardiography, which consists of rectangular (3.5 × 5 cm) or circular (5 cm) plates of German silver (nickel silver) or chrome or nickel-plated steel. Electrode jelly is rubbed on the selected site on the body and the electrode held in place at the skin site by a rubber strap or belt. This is shown in Figure 4.4(a). The modified metal plate electrode that requires no straps or adhesives for holding it in place, is the suction electrode shown in Figure 4.4(b) and these electrodes have proved to be very practical in many applications. It is used routinely to record the precordial (chest) leads of ECG, since it can be placed at a particular location and then quickly moved to the next location. It consists of a hollow, metallic, cylindrical electrode, that makes contact with the skin at its base and a rubber suction bulb which fits over its top. When the bulb is released, the suction applied against the skin holds the electrode assembly in place. Although the suction electrode is quite large, the actual contact area with the skin is relatively small resulting in a higher source impedance. Suction electrodes can, therefore, produce more distortion of the ECG, if the input impedance of the ECG amplifier is low.

A second, common variety of metal plate electrode is the metal disk shown in Figure 4.4(c). This electrode can be made of different metals (silver, platinum, stainless steel, *etc.*) and a lead wire is soldered or welded to the back surface and protected by a layer of insulating material. This electrode is mainly used as a chest electrode for recording the ECG for long-term recordings. It is maintained in place over the skin-site, applied with electrolyte paste, by a strip of surgical tape. This electrode is popularly used for surface recording of EMG or EEG, but is generally smaller in diameter (about 8 mm) than those used in recording ECG.

Pre-gelled, disposable electrodes with the adhesive already in place are now being used regularly, since these devices are ready to be applied to the patient saving a considerable amount of time. This is shown in Figure 4.4(d) and consists of a disk of plastic foam material with a silver-plated disk on one side attached to a silver-plated snap, similar to that used on clothing, in the centre of the other side. A layer of electrolyte paste covers the disk and the electrode side of the foam material is covered with an adhesive material that is compatible with the skin. A strip of release paper is placed over this side of the electrode and foam and is packed in a foil envelope to protect the water component of the paste from evaporation. To apply the electrode, the electrode packet is opened, the release paper removed and the electrode pressed on the selected skin-site on the body. Once used, the electrodes are to be disposed of.

Direct contact of the surface metal plate electrodes are a source of motion artefacts and cannot be used under dynamic conditions, that is, when a patient is moving, exercising or performing a physical task. *Floating electrodes or liquid junction electrodes* eliminate this problem.

Examples of these devices are shown in Figures 4.4(e) and 4.4(f). The principal feature is that the actual metal disk is recessed in a cavity so that it does not make direct contact with the skin. The electrical contact is established through the electrolyte paste filled in the cavity. The assembly is fixed to the prepared location on the skin with a double-sided, adhesive-tape ring. With this arrangement, the cavity does not move with respect to the metal disk and, thus, does not produce any mechanical movement of the double layer of charge. The electrode element is made of silver and is often coated with silver chloride (AgCl) as discussed earlier, in section 4.2 of this chapter.

4.2.2 Needle and Wire Electrodes

For measuring biopotentials from within the body, the transcutaneous type of electrodes are used in which the electrode or the lead wire penetrates the skin, or they may be implanted internally and connected to an implanted electronic circuit such as a radio telemetry transmitter. Figure 4.5 illustrates different types of transcutaneous needles and wire electrodes, usually made of stainless steel.

Needle electrodes are mainly used for acute measurements as they are uncomfortable for long-term implantation. The transcutaneous wire electrodes are more suitable in such situations. There are many different types of wire electrodes and techniques of introducing them through the skin. As illustrated in Figure 4.5(d), a fine wire of stainless steel ranging in diameter from 0.25 to 0.125 mm is insulated with an insulating varnish to within a few millimetres of the tip. The non-insulated tip is bent into a J-shaped structure and the tip is introduced into the lumen of the needle. After inserting the needle through the skin into the muscle at the desired location and depth, it is slowly withdrawn, leaving the electrode in place as shown in Figure 4.5(e). To remove the wire, a mild force is applied to straighten out the barb and it is pulled through the wire's tract. To provide flexibility in active muscles, the modified helical spiral electrode, as shown in Figure 4.5(f), based on the same principles as the single wire electrode is used.

4.2.3 Microelectrodes

To measure the potential difference across the cell membrane microelectrodes have to be used, which are small with respect to the cell dimensions, to avoid cellular injury. It must also be strong in order that it can penetrate the cell and be mechanically stable. The tip sizes of microelectrodes range from 0.05 to 0.10 mm. Microelectrodes can be made from solid metal needles, from metal contained within or on the surface of a glass needle, or from a glass micropipette, the details of which are described below.

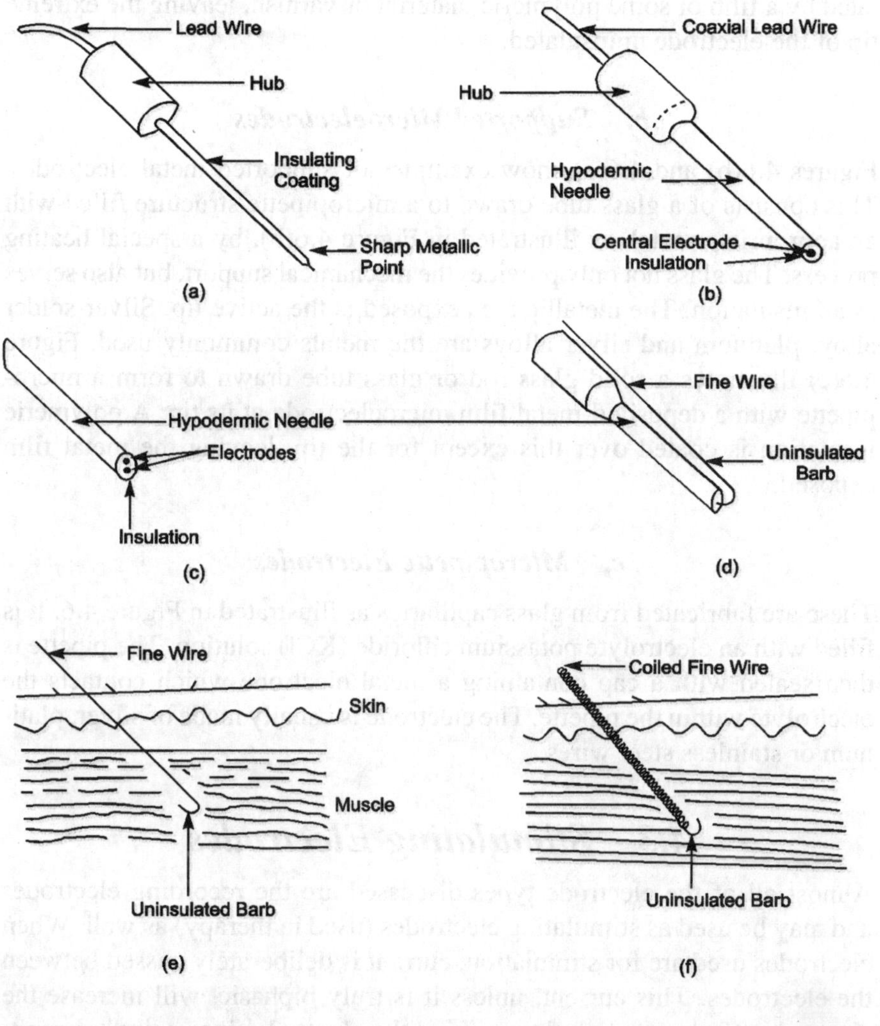

Figure 4.5 *Needle and Wire Electrodes for Intradermal Measurement of Biopotentials*

a. Metal Microelectrodes

This is illustrated in Figure 4.5(a). It consists of a fine needle made of a strong metal that is insulated up to its tip. Some suitable strong metals for such microelectrodes are stainless steel, platinum–iridium alloy and tungsten. The metal needle is supported in a large, insulated metallic shaft.

This serves as a strong metallic support to the microelectrode and also as a means of connecting it to its lead wire. The entire electrode is usually insulated by a film of some polymeric material or varnish, leaving the extreme tip of the electrode uninsulated.

b. Supported Microelectrodes

Figures 4.6(b) and 4.6(c) show examples of supported metal electrodes. This consists of a glass tube drawn to a micropipette structure filled with an appropriate metal, as illustrated in Figure 4.6(b), by a special heating process. The glass not only provides the mechanical support, but also serves as an insulation. The metallic area exposed is the active tip. Silver-solder alloy, platinum and silver alloys are the metals commonly used. Figure 4.6(c) illustrates a solid glass rod or glass tube drawn to form a micropipette with a deposited-metal film microelectrode at its tip. A polymeric insulation is coated over this except for the tip, leaving the metal film exposed.

c. Micropipette Electrodes

These are fabricated from glass capillaries as illustrated in Figure 4.6. It is filled with an electrolyte potassium chloride (KCl) solution. The pipette is then sealed with a cap containing a metal electrode which contacts the electrolyte within the pipette. The electrode is usually made of silver, platinum or stainless steel wires.

4.3 Stimulating Electrodes

Almost all of the electrode types discussed are the recording electrodes and may be used as stimulating electrodes (used in therapy) as well. When electrodes used are for stimulation, current is deliberately passed between the electrodes. This current, unless it is truly biphasic, will increase the electrode offset potential, thus making the electrodes increasingly unsuitable for future use as recording electrodes, even if silver–silver-chloride electrodes are used as stimulating electrodes. It is advisable that, once any electrode has been used as stimulating electrode, it be identified in some manner and not used thereafter as a recording electrode. Silver should be used for stimulating electrodes as it is non-toxic and does not tend to produce 'iontophoresis' due to the passage of the electric current. Iontophoresis is the driving of external ions into the skin, which can produce considerable irritation if unsuitable electrodes are used.

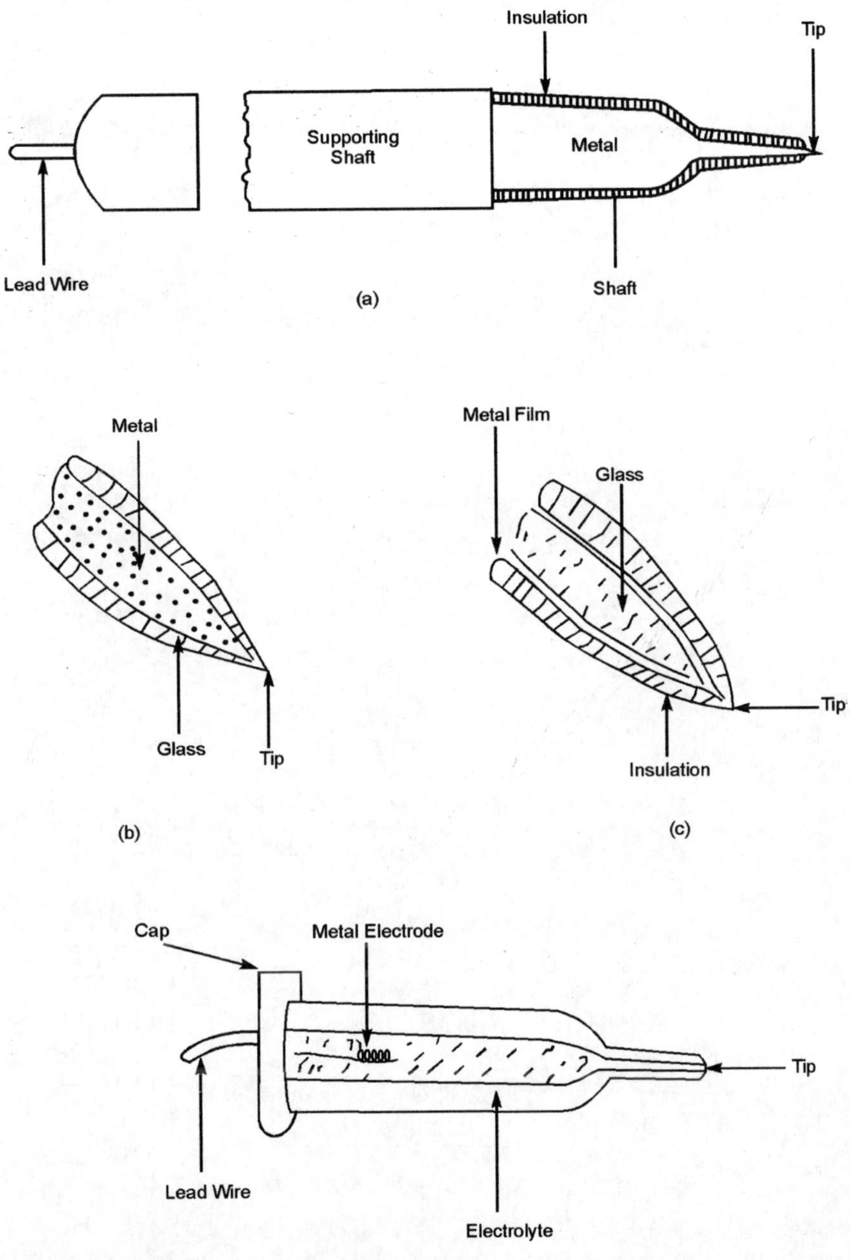

Figure 4.6 (*a*) *Metal Microelectrode for Intracellular Recordings*
(*b*) *Supported Metal-filled Glass Micropipette*
(*c*) *Supported Glass Micropipette*
(*d*) *Glass Pipette Microelectrode*

Chapter 5

Transducers for Biomedical Applications

When non-electrical biological parameters such as temperature, heart sounds, blood pressure and blood flow are to be measured or recorded with electronic equipment, the biological parameter must be made to produce an electrical signal or modify an electrical signal applied externally to the body. The devices used to convert biological parameters to electrical signals are known as *transducers* or primary sensors and the process of conversion is known as transduction. In general terms, a transducer is any device that converts energy in one form to energy in another form. However, in its applied usage, the term transducer refers to devices of a rather specialized nature. The functions of a transducer are: 1) to *sense* the presence, magnitude, change in and/or frequency of some measurand, and 2) to provide an electrical output that (when appropriately processed and applied to a read-out device) furnishes accurate, quantitative data about the *measurand*. The term measurand refers to the quantity, property, or condition which the transducer translates to an electrical signal.

5.1 Classification of Transducers

Perhaps the most satisfactory way to classify transducers is by the *electrical* principle involved in their operation. Transducers can be classified according to their application, based primarily on the physical quantity, property or condition that is measured. A sharp distinction between, and classification of, types of transducers is difficult. Table 5.1 shows a classification of transducers according to the electrical principles involved. The first part of the table lists the *passive* transducers. These transducers require an external power and their output is a measure of some variation, such as resistance inductance, capacitance, *etc*. In the second category are the *active* or *self-generating* transducers. These transducers do not require any external power and they produce an analog voltage or current when stimulated by some physical form of energy.

Table 5.1 Some Types of Transducers and Applications

Type	*Principle of Operation*	*Application*
1) Displacement		
Variable Resistance	Displacement moves a contact slider on a resistance	Limb movement, Joint flexion (goniometer)
Strain Gauge i) Wire ii) Semiconductor	Deformation leads to change in resistance	Blood vessel distension
Linear Variable Differential Transformer (LVDT)	Inductive coupling on two coils change with displacement	Limb movement
Capacitance	Variation in separation of plates of a capacitor produces change in capacitance	Ballistocardiography
Optical	Light intensity change with displacement	Digital plethysmography
	Separation between two light beams varying with movement	Surgical robotics
Magnetostriction	Movement of magnetic field produces variation in effective electrical resistance of material	Finger movement
Ultrasonic	Time interval between emitted ultrasonic pulse and pulse received back after reflection is proportional to the distance between emitter and reflector	Brain midline detection
2) Velocity		
Potentiometer	Time differential of voltage from a linear/circular resistance proportional to linear/angular velocity	Joint motion
Electromagnetic	Potential generated across a wire moving in a magnetic field proportional to velocity	Gait analysis
Ultrasonic	Doppler frequency shift between emitted and received wave related to velocity	Heart-wall motion

(contd...)

(...continues)

Type	*Principle of Operation*	*Application*
3) Acceleration		
Strain Gauge	A displacement sensor affixed to an inertial platform	Hand movement
Piezoelectric	Time differential of voltage generated related to acceleration	Muscle contraction
Electromagnetic	Time differential of electromagnetic induction, EMF produced proportional to acceleration	Gait studies
4) Flow		
Hot Wire	Cooling of heated wire related to fluid flow	Respiratory gas flow
Venturi	Bernoulli's principle gives a pressure differential flow expression	Blood flow in heart-lung machines
Turbine	Rotational speed of turbine blade related to fluid flow rate	Dialysate flow in artificial kidney
Electromagnetic	For ionic liquids moving in a magnetic field, EMF generated proportional to the flow rate	Blood flow in arteries
5) Force		
Strain Gauge	Displacement transducer combined with a spring device	Foot pressure
LVDT	Displacement transducer combined with a spring device	Grip force
6) Pressure		
Strain Gauge	Displacement transducer coupled to a resilient diaphragm	Venous blood pressure
LVDT	Displacement transducer coupled to a resilient diaphragm	Arterial blood pressure
Piezoresistor	Displacement transducer coupled to a resilient diaphragm	Finger contact pressure
7) Vibration		
Piezoelectric	Alternating compression/ extension, generates EMF in synchrony with the movement	Hand tremor
Electromagnetic	Voltage induced in a wire linked to the vibrator and	Pulse transducer

(contd...)

(...continues)

Type	*Principle of Operation*	*Application*
	placed in a magnetic field, has same frequency as that of the vibration.	
8) Temperature		
Thermocouple	Peltier effect generates EMF when temperature difference exists between 'hot' and 'cool' junctions of dissimilar metals	Skin contact thermometry
Thermistor	Resistance of special materials increase or decrease with an increase in temperature	Oral temperature
9) Humidity		
Resistance	Ion mobility change with water concentration leads to change in conductivity	Exhaled gas moisture level
Deformation Hygrometer	Mechanical stresses induced in material on absorption of moisture	Hospital environment monitoring
10) Illumination		
Photovoltaic cell	Charge carrier energy level shift with illumination leads to generation of EMF	Colorimetry
Photoresistive (Light Dependent Resistor-LDR)	Charge carrier mobility change induces variation in material conductivity	Photo-plethysmograph
11) Chemical		
Glass electrode	Selective permeability of glass gives potential difference with differential hydrogen ion concentration (pH)	pH measurement
Ion-selective electrode	In one form, enzyme immobilized on the electrode gives reaction products which alter electrical status of the electrode	Serum K^+ estimation
Ion-selective field effect transistor	Combines the electric properties of gate-insulator field effect transistor with electrochemical properties of ion-selective electrodes	Ca^{++} estimation
Thick-film sensors	Chemically active, ion-specific layer of polyvinyl chloride	Na^+

(contd...)

(...continues)

Type	*Principle of Operation*	*Application*
	(PVC) or other material patterned to carry out multiple assays with a single device	
12) Biosensor		
Enzymatic sensors	Enzyme reactions release charged species which are estimated by techniques such as conductimetry, amperometry.	Blood glucose/ urea estimation
Whole cell sensor	Recombinant forms of whole cells, such as *E. coli,* produce reactants in combination with the measurand to be further assessed by electrochemical sensors.	Biological oxygen demand

5.2 Selecting a Transducer

The transducer, or sensor, as it is sometimes called, has to be physically compatible with its intended application. In the selection of a transducer, there are approximately eight areas of concentration:

1. Operating range: Chosen to maintain range requirements and good resolution.
2. Sensitivity: Chosen to allow sufficient output.
3. Frequency response and resonant frequency over the needed range: Will the resonant frequency be excited?
4. Environmental Compatibility: Temperature range, corrosive fluids, pressures, shocks, interaction, size, and mounting restrictions.
5. Minimum sensitivity: To expected stimulus other than the measurand.
6. Accuracy: Repeatability and calibration errors as well as errors anticipated from sensitivity to other stimuli.
7. Usage and ruggedness of both mechanical and electrical intensities versus size and weight: Who will be installing and using the transducer?
8. Electrical: Length and type of cable required; the signal-to-noise ratios when combined with an amplifier and frequency response limitations.

5.3 General features of a Biomedical Instrumentation System

Figure 5.1 shows the general schematic layout of a biomedical instrumentation system. The physiological parameter is detected by a suitable *input transducer;* if the physiological event is electrical in nature, such as a biopotential from an organ (say, ECG/EMG), then suitable electrodes are used for detecting the bioelectric event. The transduced signal which is

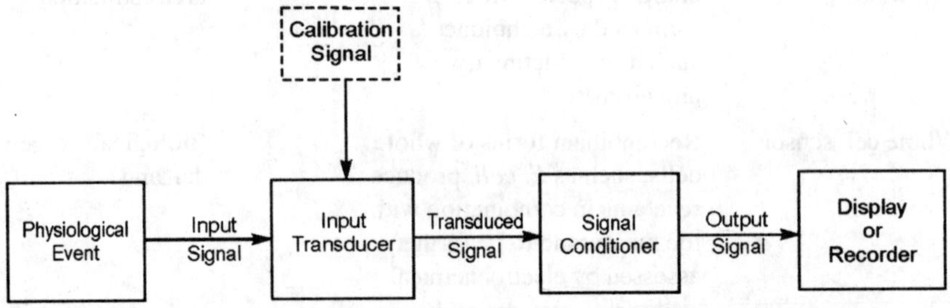

Figure 5.1 *General Schematic Layout of Biomedical Instrumentation*

proportional to the input signal is fed to the signal conditioner, which amplifies and modifies the transduced signal to match with the final stage, consisting of a display unit or chart recorder calibrated directly in terms of the input signal amplitude. It will be necessary to use some form of signal-transmission system, like transmission through wires or a telemetry system, to link the signal conditioner and the final stage of display or the recorder.

This chapter should not be regarded as a general reference on transducers; it specifically discusses (only some) transducers from the standpoint of biophysical measurement.

5.4 Variable Resistance Transducers

Transducers using variable resistance occur in two forms: a) wiping contact on the resistive track, b) deformation-sensitive resistive element (strain gauge).

5.4.1 Resistive Track Transducers

In this type of transducer, the wiping contact may either move round a circular resistive track or linearly along a straight track. For transducing relatively large and slowly occurring position changes, this is the simplest transducer. The circular resistance type is especially suitable for use with a circular motion. Such devices are readily available in the form of the ordinary radio potentiometer and, in refined forms, are capable of the highest accuracy of any type of transducer. Linear variable resistance will usually have to be constructed for a particular application. They may be applied directly to the measurement of linear movement. Examples of these are shown in Figure 5.2. These potentiometers have the advantage that their resistance can be changed from zero to the maximum resistance up to several

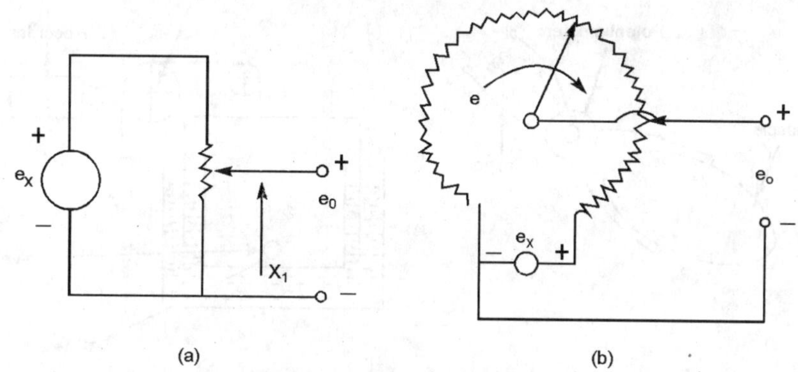

Figure 5.2 *Potentiometer for Converting (a) Linear (b) Angular Displacement to Electrical Signals*

mega ohms. The circuitry associated with them is quite simple, as shown in Figure 5.3.

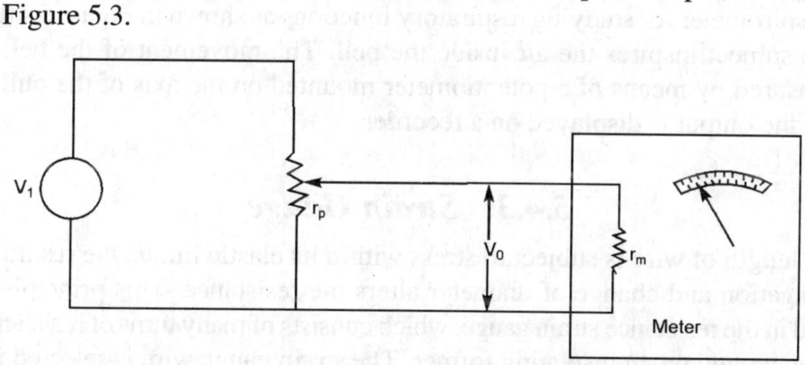

Figure 5.3 *Circuit for Potentiometer Transduction*

5.4.2 *Application of Resistive Potentiometers*

Although there are few physiological applications of this type of transducer at present, the unusually high efficiency attainable will no doubt encourage wider applications. Its manipulation is for the measurement of displacement.

Respiration can be detected by measuring changes in thoracic (chest) circumference, by connecting a rotary potentiometer to a chest band on a human being. It was used earlier to measure the contractions of a skeletal muscle. In this transducer, the potentiometer serves as the pivot for the caliper arms, which embraces the belly of the muscle as shown in Figure 5.4(a). The contraction of the muscle causes the arm to be driven apart, and its motion is measured by the potentiometer. It is widely used in

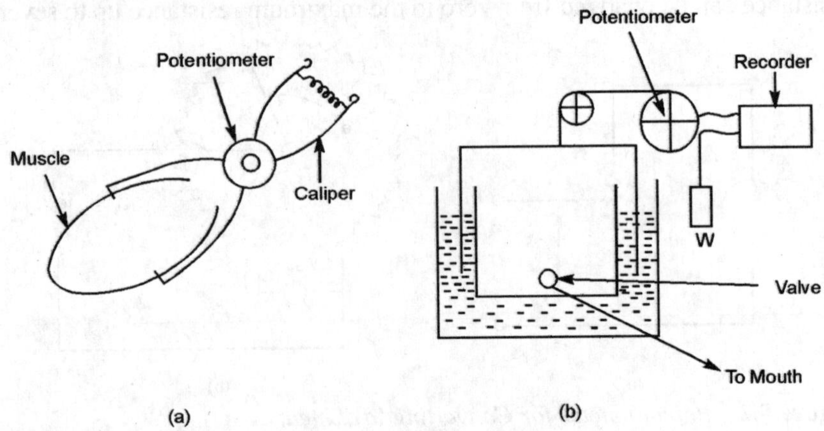

Figure 5.4 *Potentiometer Transducer Applications in (a) A Caliper Myograph (b) A Spirometer*

the spirometer for studying respiratory functions as shown in Figure 5.4(b). The subject inspires the air inside the bell. The movement of the bell is measured by means of a potentiometer mounted on the axis of the pulley and the output is displayed on a recorder.

5.4.3 *Strain Gauge*

If a length of wire is subject to stress within its elastic limits, the resulting elongation and change of diameter alters the resistance. This principle is used in the resistance strain gauge, which consists of many turns of resistance wire, wound on an insulating former. The strain gauge wire is selected for the minimum temperature coefficient of resistivity. Constantan, nichrome V and other similar materials are used for this purpose. They may be used to measure extremely small displacements, of the order of *nanometers.*

For a wire of cross-sectional area a (m^2), resistivity ρ (ohm-m) and length L (m), the resistance is given by

$$R = \frac{\rho L}{A} \text{(R will thus change if L and A change)} \qquad ....(5.1)$$

Differentiating on both sides

$$dR = \frac{\rho}{A}dL + \frac{L}{A}d\rho - \frac{\rho L}{A^2}dA \quad ....(5.2)$$

dividing (5.2) by (5.1), we get the fractional resistance change as

$$\frac{\Delta R}{R} = \frac{\Delta L}{L} = \frac{\Delta A}{A} + \frac{\Delta\rho}{\rho} \quad ....(5.3)$$

Poission's ratio μ defined as the ratio of strain in the lateral direction to strain in the axial direction. Therefore,

$$\mu = -\frac{\Delta d/d}{\Delta L/L} \quad ....(5.4)$$

Since $A = \pi d^2/4$, so that $\Delta A/A = 2\Delta d/d$, Eq. (5.4) can be expressed as

$$\Delta A/A = 2\mu\Delta L/L \quad ....(5.5)$$

Substituting Eq. (5.5) into Eq. (5.3), we get

$$\frac{\Delta R}{R} = \underbrace{(1+2\mu)\frac{\Delta L}{L}}_{\text{Dimensional effect}} + \underbrace{\frac{\Delta\rho}{\rho}}_{\text{Piezoresistive effect}} \quad ....(5.6)$$

Hence, the change in resistance can be expressed as the sum of two terms—the first due to simple dimensional changes and the second due to change in resistivity with strain, commonly known as **piezoresistive effect.** The *gauge factor* G or strain sensitivity of gauge is defined as the fractional change in resistance divided by the fractional change in length; that is

$$G = \frac{\Delta R/R}{\Delta L/L} = (1+2\mu) + \frac{\Delta\rho/\rho}{\Delta L/L} \quad ....(5.7)$$

Equation (5.7) provides a useful means for comparing the performance of various gauge materials. For most metals, the Poisson's ratio is approximately 0.3, so that the Gauge Factor becomes

$$G = 1.6 + \frac{\Delta\rho/\rho}{\Delta L/L} \quad ....(5.8)$$

The gauge factor of most metals is approximately 2.0, whereas that of silicon is 60 times larger. The materials of strain gauge are Advance (50% Cu and 45% Ni) and Isoelastic (36% Ni, 8% Cr, 4% Mn, *etc.*) which have gauge factors of 2.1 and 3.5 to 3.6, respectively.

Strain gauges are classified as Bonded and Unbonded types. The bonded type has the wire mounted on a paper backing as shown in Figure 5.5(a) and the paper could be pasted to the surface of the body under strain. The unbonded type is simply wire-mounted between two supports which move with respect to each other.

In most bonded strain gauges, a length of strain gauge wire of very small diameter is mounted in a zig-zag fashion on a piece of paper which, in turn, is mounted on a surface whose elongation is to be measured. Unfortunately, the maximum change of resistance which can be obtained before the gauge wires pass their elastic limit, and hence change their calibration, is about 4%. Furthermore, the power dissipated in the gauge must not exceed, typically, about 25 mW so that its characteristics may not be affected. For this reason, the strain gauge is most applicable for displacement measurement of less than a few thousandths of a centimeter. Due to their sensitivity to humidity, unbonded strain gauges have limited applications in biological research.

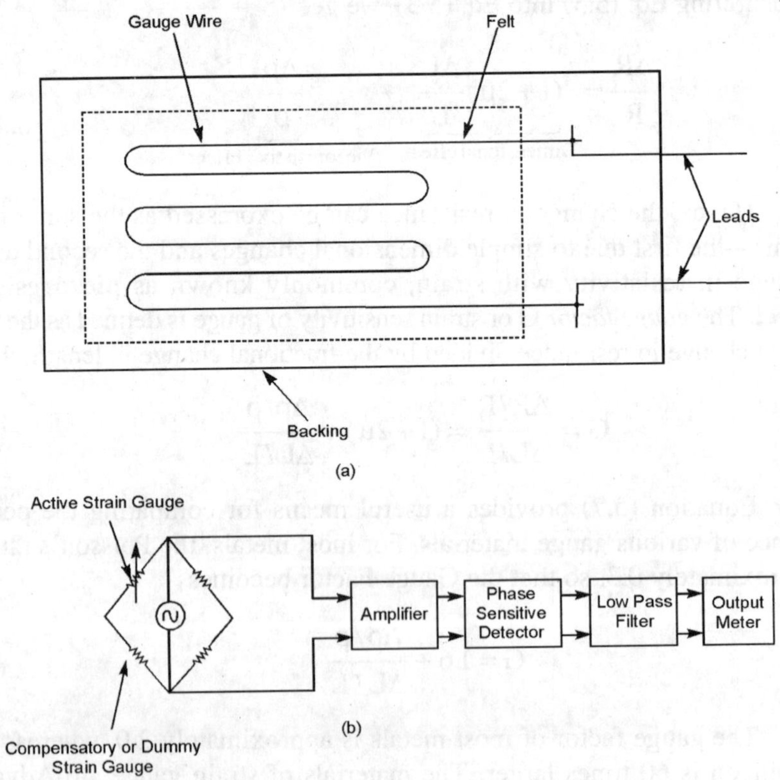

Figure 5.5 *(a) Bonded Wire Strain Gauge*
(b) Strain Gauge Measurement Scheme

In order to convert the variations in resistance of strain gauges into recordable electrical signals, the strain gauge is used as one arm of a wheatstone bridge. Because temperature errors may be encountered, the use of a single strain gauge element is uncommon. To reduce temperature errors, pairs of strain gauges are usually employed in a bridge circuit. A simple method of temperature compensation is the use of a dummy gauge in the bridge circuit. The dummy gauge (of equal ohmic value as an active gauge) is mounted in the adjacent arm of the active gauge, thus cancelling output differentials due to temperature variations of the resistance of the gauges. It is possible to commercially obtain, two strain gauges mounted with their axes perpendicular to each other. These are mounted to the surface under study, with one used as an active gauge while the other is used for temperature compensation. The output for such a strain gauge is of the order of a few millivolts for full-scale elongation, when the bridge is powered with 6 or 12 V DC. An alternative to the use of DC wheatstone bridges is the use of an AC power source operating at several KHz and the incorporation of the strain gauge into an AC bridge. In this case, the output of the bridge is an alternating voltage which is easier to amplify and, subsequently, to record. The bridge circuit is shown in Figure 5.5(b).

5.4.4 Semiconductor Strain Gauges

Recently, the semiconductor materials (silicon and germanium) have been employed in transducers. Because of the small size of the device, the mass is low. They have a large gauge factor, ranging from 50 to 200. They are, however, sensitive to temperature fluctuations and often behave in a non-linear manner. This strain gauge microtransducer can be fixed at the tip of a catheter.

5.4.5 Applications of Strain Gauge Transducers

The strain gauge is very frequently used in biomedical studies, some of which are enumerated below.

1. By mounting it suitably on the walls of the cardiac muscles, the force of the contraction of the cardiac muscles can be measured continuously.
2. Blood pressure inside the heart or blood vessels is an important parameter, useful for diagnosing cardiovascular abnormalities. The strain gauge is mounted on the tip of a catheter which is inserted into the heart through a vein. In front of the strain gauge, is mounted a diaphragm, the deflection of which varies with the blood pressure and, in turn, alters the resistance of the strain gauge.

3. It is used in the volumetric differential low pressure transducer. It consists of two cavities separated by a large sensitive diaphragm. When the diaphragm is displaced, it activates a cantilever on which the strain gauges are mounted. The strain gauges transduce the movement into changes of electrical resistance. This transducer is primarily used for measuring small volume changes, but it may be used to measure pressure up to 5 mm Hg.
4. The respiratory rate of flow may be accurately recorded using this transducer as a differential transducer to measure the drop in pressure across a fine mesh screen. There is a linear relation between the rate of flow and the pressure drop across the screen.
5. Strain gauges are widely employed for sensing the loading (pressure) while fitting artificial limbs.

In general, resistance transducers have the advantage of linearity of response, freedom from hysteresis and effective temperature compensation.

5.5 *Variable Inductance Transducers*

The differential transformer is a passive inductive transducer. It is also known as the linear variable differential transformer (LVDT). It consists of three coils: a primary and two identical secondaries, as shown in Figure 5.6. The primary (central coil of LVDT) is excited by an alternating current which produces a field that induces equal voltages in the two secondaries (adjacent coils). The secondaries are connected in series opposition,

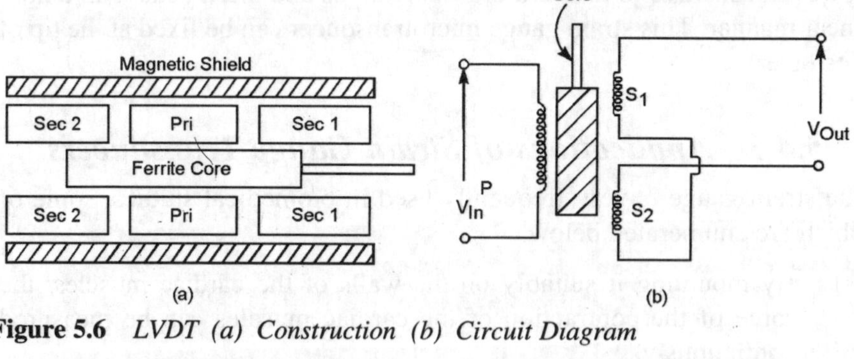

Figure 5.6 *LVDT (a) Construction (b) Circuit Diagram*

so that the voltage generated in one cancels that from the other. The insertion of a core imbalances the system, so that the voltage generated in the secondaries are no longer equal. The relationship between the output voltage and the position of the core is shown in Figure 5.7. With the core in the mid-position, the output at the secondary terminals is zero. An output of the same magnitude is produced if the core is displaced by an equal amount in either direction from its mid-position. The phase of the output differs, however, by 180° on either side of the neutral position.

In demodulating the output, it is necessary to preserve this 180° phase-shift that occurs in going through the null, since it is the only means of

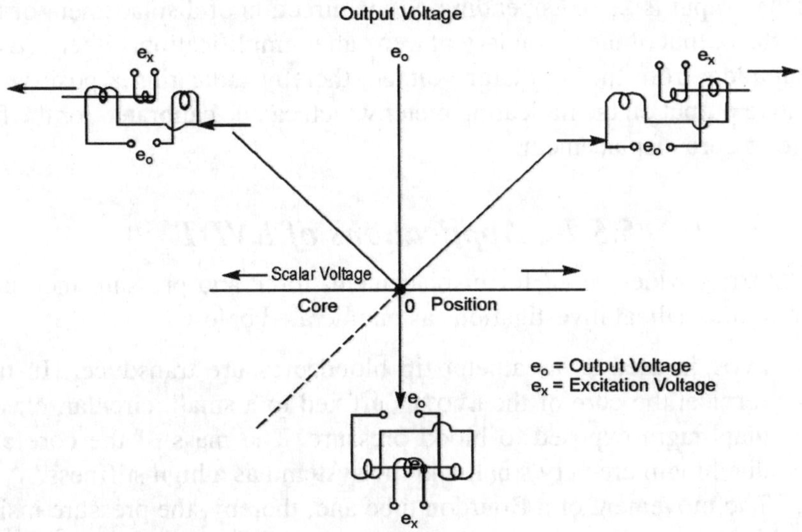

Figure 5.7 *LVDT Output Versus Core Position*

determining the sign of the displacement. If a simple diode detection is used at the output, the output will have the same polarity for both positive and negative displacement of the core. Therefore, to determine the direction of displacement of the core, a phase sensitive detector is used as shown in Figure 5.8 which yields a positive or negative output according to the

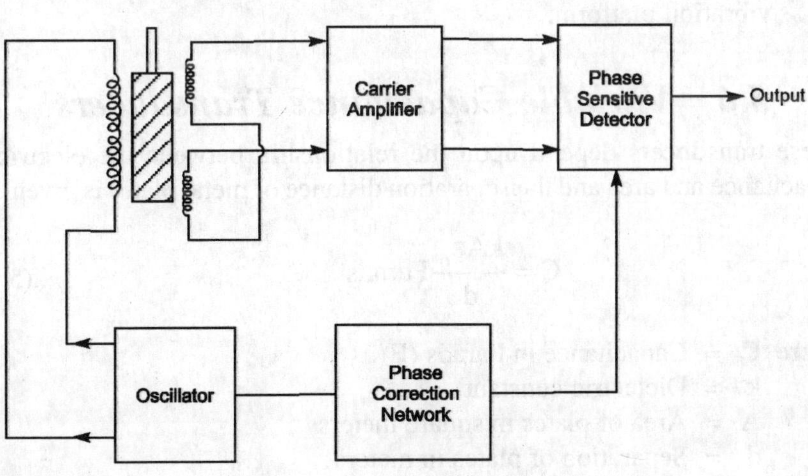

Figure 5.8 *Phase Sensitive Detection System for LVDT*

direction of displacement with respect to the null. In this scheme, the oscillator voltage (excitation voltage) and that derived from the LVDT are added before rectification. With the core in the central (null) position, the oscillator voltage corrected for phase shifts is fed to the phase sensitive detector so that the output is zero. Depending on the direction of displacement of the core, the output of the secondary of LVDT after amplification, either adds to or subtracts from the oscillator voltage, thereby indicating a positive or negative output on the indicating meter which can be calibrated for the full range of core displacement.

5.5.1 Applications of LVDT

The LVDT is widely used for displacement, force and pressure measurement in biomedical investigations as enumerated below.

1. LVDT is used as a catheter-tip blood pressure transducer. In this service, the core of the LVDT is affixed to a small, circular, elastic diaphragm exposed to blood pressure. The mass of the core and diaphragm are very small and the system has a high stiffness.
2. The movement of a Bourdon tube and, thereby, the pressure inside the tube can be measured by connecting the core of the LVDT to the tip of the Bourdon tube.
3. LVDT is used in Ballistocardiography for picking up the movement of the ballistocardiograph platform (on which a subject lies supine) due to the pumping action of the heart.
4. Vibrations of several body segments can be monitored by using LVDT to study the effects of vibrations on the human body placed on a vibration platform.

5.6 Variable Capacitance Transducers

These transducers depend upon the relationship between the electrical capacitance and area and the separation distance of metal plates is given by

$$C = \frac{kA\varepsilon_o}{d} \text{ Farads} \qquad \text{....(5.9)}$$

where C = Capacitance in Farads (F)
k = Dielectric constant
A = Area of plates in square meters
d = Separation of plates in meters
ε_o = 8.854×10^{-12} F/m

The capacitance thus varies, if the distance between the plates or the area of the plates is changed. Figure 5.9 shows a number of transducers based on this principle.

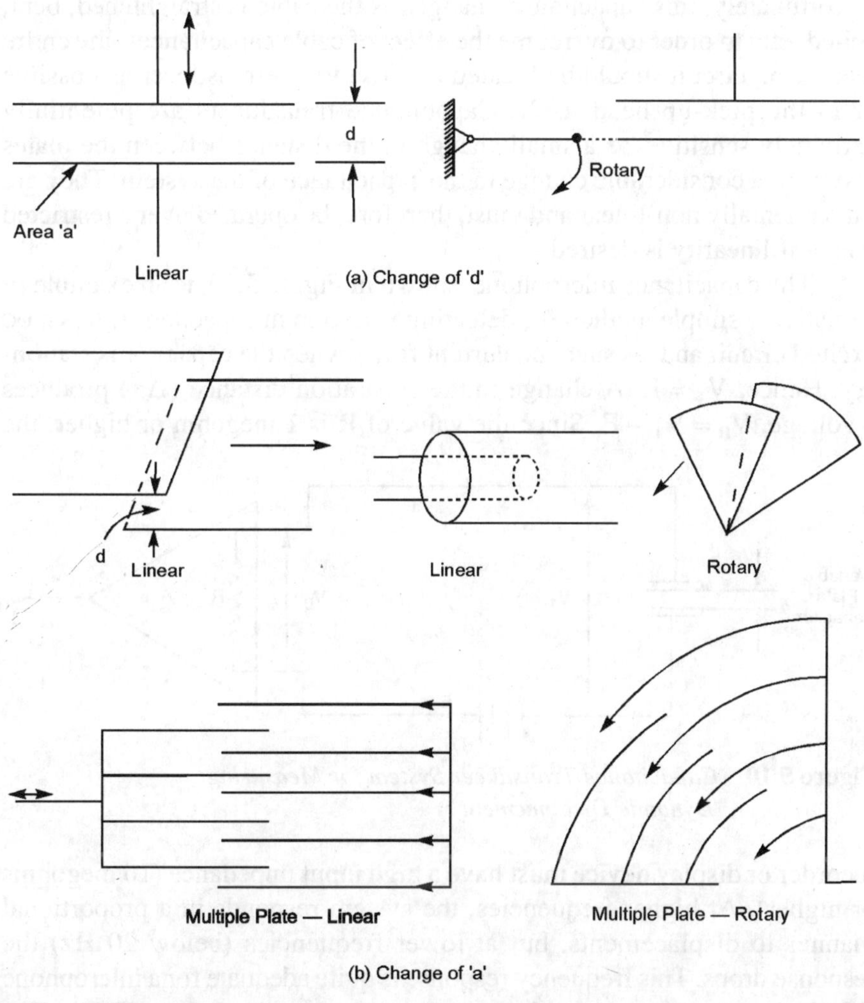

Figure 5.9 *Capacitance Displacement Transducers*

This change of capacitance has to be converted into an electrical signal which can be rectified, amplified and recorded. This is achieved by so incorporating this capacitance into a circuit that determines the frequency of oscillation of an oscillator, with an output tank circuited and tuned to a fixed frequency. The output signal varies with this change in the frequency.

Since the capacitance transducer circuits are very sensitive, the transducer must be well designed for temperature compensation and thermally insulated to prevent temperature changes. Such a design has frequently turned out to be difficult to achieve. Secondly, the cables connecting the transducer to the measuring circuit themselves have large capacitance. Unfortunately, this capacitance changes as the cable is straightened, bent, coiled, etc. In order to overcome the effect of cable capacitances, the entire measuring circuit should be located as close to the transducer as possible or in the pick-up head itself. Capacitance transducers are potentially extremely sensitive, as a small change in the distance between the plates results in a considerable change in the capacitance of the system. They are fundamentally non-linear and must, therefore, be operated over a restricted range, if linearity is desired.

The capacitance microphone, shown in Figure 5.10, is an example of a relatively simple method for detecting variation in capacitance. It is a DC excited circuit and, as such, no current flows when the capacitor is stationary. Hence, $V_1 = E$. A change in the separation distance (Δx) produces a voltage, $V_0 = V_1 - E$. Since the value of R is 1 megohm or higher, the

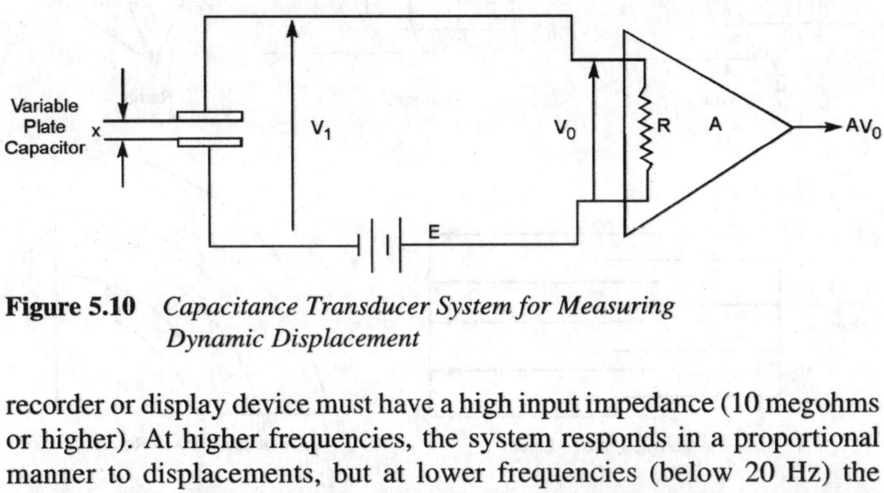

Figure 5.10 *Capacitance Transducer System for Measuring Dynamic Displacement*

recorder or display device must have a high input impedance (10 megohms or higher). At higher frequencies, the system responds in a proportional manner to displacements, but at lower frequencies (below 20 Hz) the response drops. This frequency response is quite adequate for a microphone that does not measure sound pressures at frequencies below 20 Hz. The lack of a DC response is, however, a serious disadvantage for measuring most of the other physiological variables.

5.6.1 Applications of Capacitive Transducers

1. The capacitance method has often been applied to the measurement of physiological events, particularly blood pressure. An elastic element exposed to blood pressure constitutes one plate of the

capacitor, the other plate is nearby and fixed. To obtain a rapid response time, the elastic element is made as small and as stiff as possible.

2. The capacitance microphone is used for the detection of heart sounds.
3. A miniature capacitance microphone is used as a high fidelity pulse pick-up.
4. An unusual application of the capacitance change principle employs the dielectric property of the living tissues itself as part of the capacitor. This principle is used by placing electrodes on the chest and back and correlating the output of this, to record volume and cardiac output for each heartbeat.

5.7 Thermoresistive Transducers

The change in resistance of certain metals and semiconductors constitutes the basis of thermoresistive temperature transducers. The temperature coefficient of metals is normally positive, corresponding to an increase in resistance with temperature. On the other hand, semiconductors made of ceramic materials are thermoresistive, with a high negative temperature coefficient.

A type of resistance thermometer of special interest for biomedical applications is the thermistor, in which the resistive element is a semiconductor with a very high negative temperature coefficient of resistivity. Thermistors are made in a wide variety of forms, such as wafers suitable for applying to the skin surface, rods which can be used for rectal, oral or similar insertions and tiny beads so small (less then 0.5 mm in diameter) that they can be mounted at the tip of a hypodermic needle for insertion into the tissues. The glass-encapsulated lead thermistor is the one most commonly used in biomedical applications. The glass coating protects the sensing element from the hostile environment of the body without

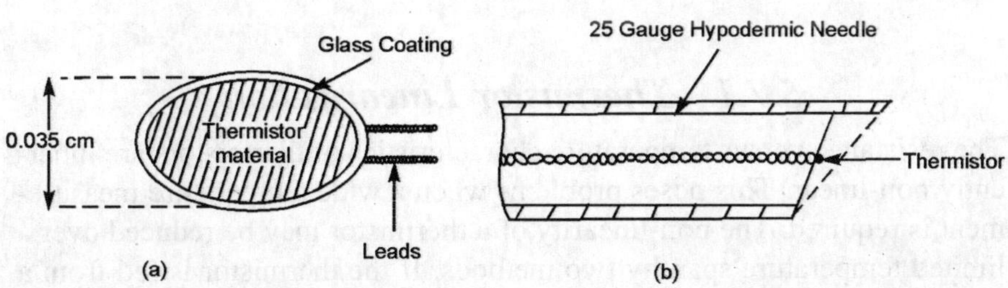

Figure 5.11 *Examples of Thermistor Probes: (a) Bead Thermistor (b) Bead Thermistor Mounted at the End of a Hypodermic Needle*

significantly affecting the thermal response time of the system which is very small. Examples of lead thermistor probes are shown in Figure 5.11. The resistivity of thermistor semiconductors which are used for biomedical applications lies between 0.1 to 100 ohm-m.

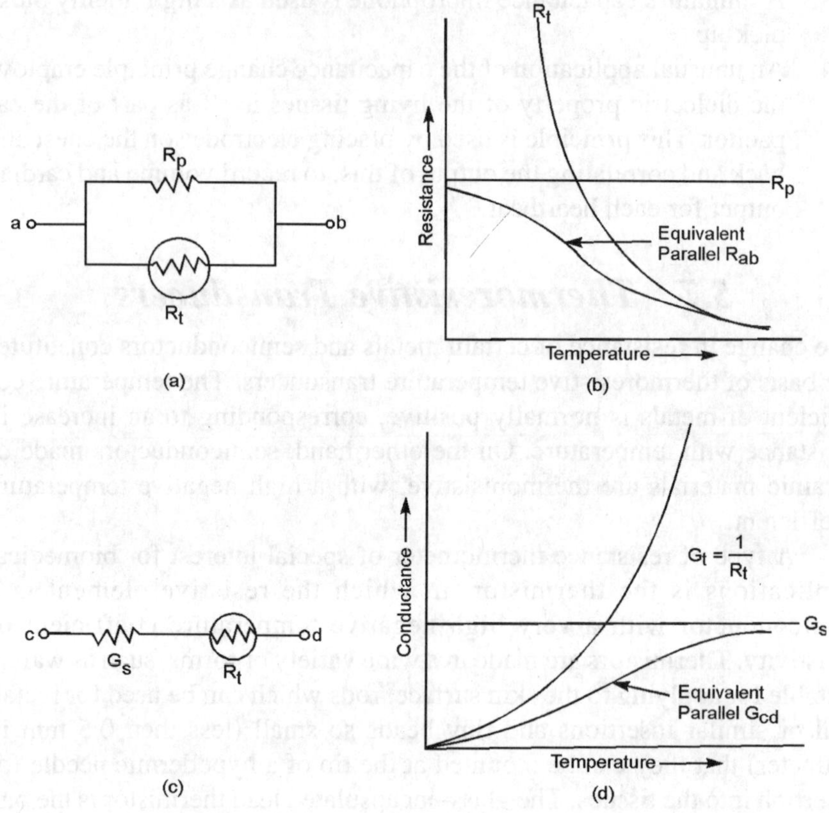

Figure 5.12 *(a) Thermistor Linearization and (b) Shunt Network Giving Improved Resistance-Temperature Linearity (c) and (d) Series Network Giving Improved Conductance-Temperature Linearity*

5.7.1 Thermistor Linearization

The resistance versus temperature characteristics of thermistors are inherently non-linear. This poses problems when a wide temperature measurement is required. The non-linearity of a thermistor may be reduced over a limited temperature span by two methods. If the thermistor is fed from a constant-current source and the voltage across the thermistor is used to indicate the temperature, linearization can be achieved by shunting the thermistor with an appropriately chosen resistor R as shown in Figure 5.12(a), and the plot of resistance versus temperature for the total resistance with or without linearization is shown in Figure 5.12(b). Alternatively, when the current through the thermistor for a fixed, applied emf is used to indicate the temperature, the series arrangement shown in Figure 5.12(c) may be used. The conductance versus temperature characteristic with and without the series resistance is shown in Figure 5.12(d).

The circuitry used for the thermistor read-out is essentially the same as for conductive sensors, and many of the same techniques like bridge circuits and operational amplifier circuits apply. A DC differential temperature bridge using matched thermistors R and R of the value of 100 kilo-ohms (± 1%) is shown in Figure 5.13. The DC stability of this bridge is not normally a problem, since the output voltage of the bridge – even for the temperature differences of 0.01 % – is larger than the DC drift of good integrated circuit operational amplifiers. Like other resistance

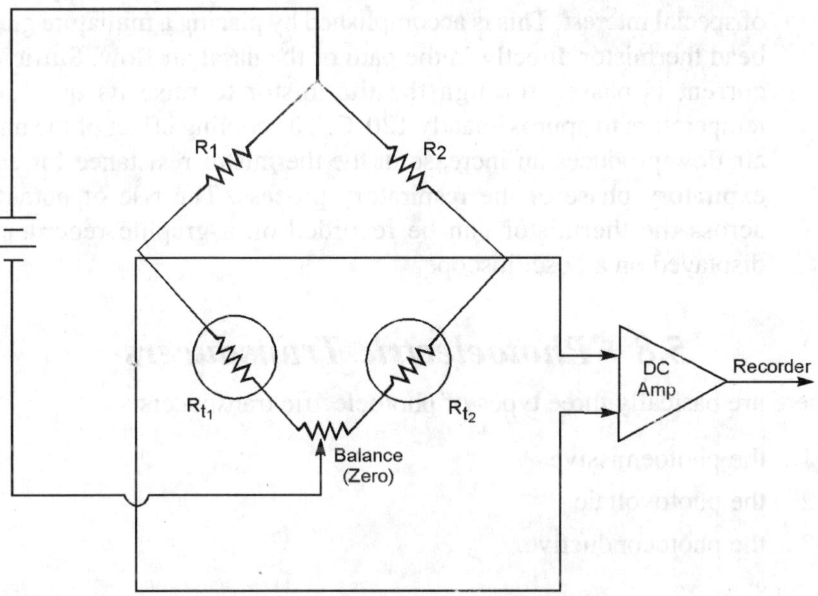

Figure 5.13 *DC Differential-Temperature Bridge*

thermometers, measurement involves passing a current through them with the result that they dissipate power and heat themselves. Their power dissipation must be limited to something of the order 1/10 mW and, because of the great change in the resistance over a reasonable temperature excursion, they are fed from a constant current source. The thermistor also has some hysteresis, if subjected to rapid large temperature fluctuation, but these will not seriously affect the normal biomedical work. In general, the instrument should not be subjected to a greater temperature gradient than 10°C/min or to a overall excursion more than 50–60°C without recalibration. Under the best conditions, a measurement to 0.01°C may be obtained.

5.7.2 Application of Thermoresistive Transducers

1. They are used for continuous measurement of skin and body core temperatures.
2. They are used for the measurement of blood flow. A heated thermistor is mounted on the tip of a catheter or hypodermic needle which is inserted into the blood vessel. The cooling effect, which is proportional to the blood flow, alters the resistance of the thermistor which is related to the blood flow.
3. The use of the thermistor for the measurement of respiratory rate is of special interest. This is accomplished by placing a miniature glass-bead thermistor directly in the path of the nasal air flow. Sufficient current is passed through the thermistor to raise its quiescent temperature to approximately 120°C. The cooling effect of the nasal air flow produces an increase in the thermistor resistance for each expiratory phase of the respiratory process. The rise of potential across the thermistor can be recorded on a graphic recorder or displayed on an oscilloscope.

5.8 Photoelectric Transducers

There are basically three types of photoelectric transducers:

1. the photoemissive
2. the photovoltaic
3. the photoconductive.

5.8.1 Photoemissive Tubes

The photoemissive tube consists of an evacuated or gas filled bulb with two electrodes—one, the cathode on which is a coating of specially prepared material that releases electrons when illuminated, and the other, the anode. The coating materials usually employed are antimony, silver, and bismuth—each type exhibiting its own special characteristics. The electron flow released by the light quanta constitute a current which is linearly proportional to the intensity of the incident light. In the vacuum type, the current produced is small, which needs amplification to operate an indicator.

When higher currents are required, gas-filled tubes are used. By this arrangement, the primary electrons released by the incident light collide with the gas molecules and produce secondary electrons, thus increasing the available current.

Both vacuum- and gas-filled photoemissive tubes respond quickly to changes in light intensity. The spectral sensitivity of these tubes seldom extends below 200 mμ and most tubes are sensitive to visible light.

By incorporating additional anodes (dynodes) each at the higher potential, a current amplification can be obtained through secondary electron emission. Electrons emitted from the photoelectric surface strike the first dynode, releasing secondary electrons which are accelerated towards

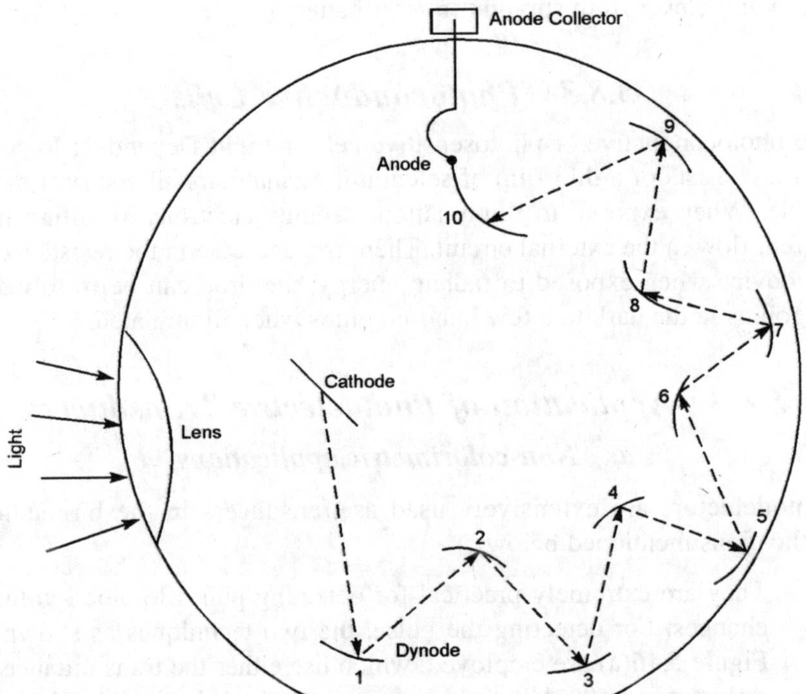

Figure 5.14 *A Photomultiplier*

a second dynode at a higher potential where the process is repeated. With a chain of nine dynodes, ending finally in an anode, multiplications of up to 10^6 can be obtained. Such a device is called a photomultiplier, which is shown in Figure 5.14.

5.8.2 Photovoltaic Cell

These cells consist of a thin coating of selenium on an iron or steel backing. A thin, transparent film of metal is coated over the selenium. The film and selenium are insulated from each other, and this constitutes the barrier layer, which, when illuminated, releases electrons, thus causing a potential difference across the barrier. The transparent metal film becomes negative and the selenium becomes positive. So, this cell comes under the category of active transducers.

Selenium photo cells are sensitive to the visible spectrum (300–700 mμ) with a peak around 550–570 mμ. The silicon photovoltaic cell is a P-N junction device which produces large currents and is used for producing power from sunlight (a solar battery).

5.8.3 Photoconductive Cells

The photoconductive or photosensitive cell or Light Dependent Resistor (LDR), consists of a thin film of selenium, germanium, silicon or a metal halide. When exposed to illumination, it emits electrons, resulting in a current flow in the external circuit. There is a decrease in the resistance of the device when exposed to radiant energy, the drop can be from many megohms in the dark to a few hundred ohms when illuminated.

5.8.4 Application of Photoelectric Transducers

a. Non-colorimetric applications

Photodetectors are extensively used as transducers in the biomedical applications mentioned below.

i. They are extremely practical for detecting pulsatile blood-volume changes. For detecting the pulse, the two techniques as shown in Figure 5.15(a) are employed, which use either the transmittance or reflectance method. In the first technique, the pulsating blood modulates the optical density, and in the second, the pulse alters the amount of reflected light. Although this method does not indicate the total flow, it produces a signal which immediately shows the changes in blood flow. By employing suitable methods, the pulse rate can be counted.

ii. The respiratory volume can be measured by measuring changes in the chest circumference with a pneumograph which has a photodetector incorporated in it, as shown in Figure 5.15(b). The closed

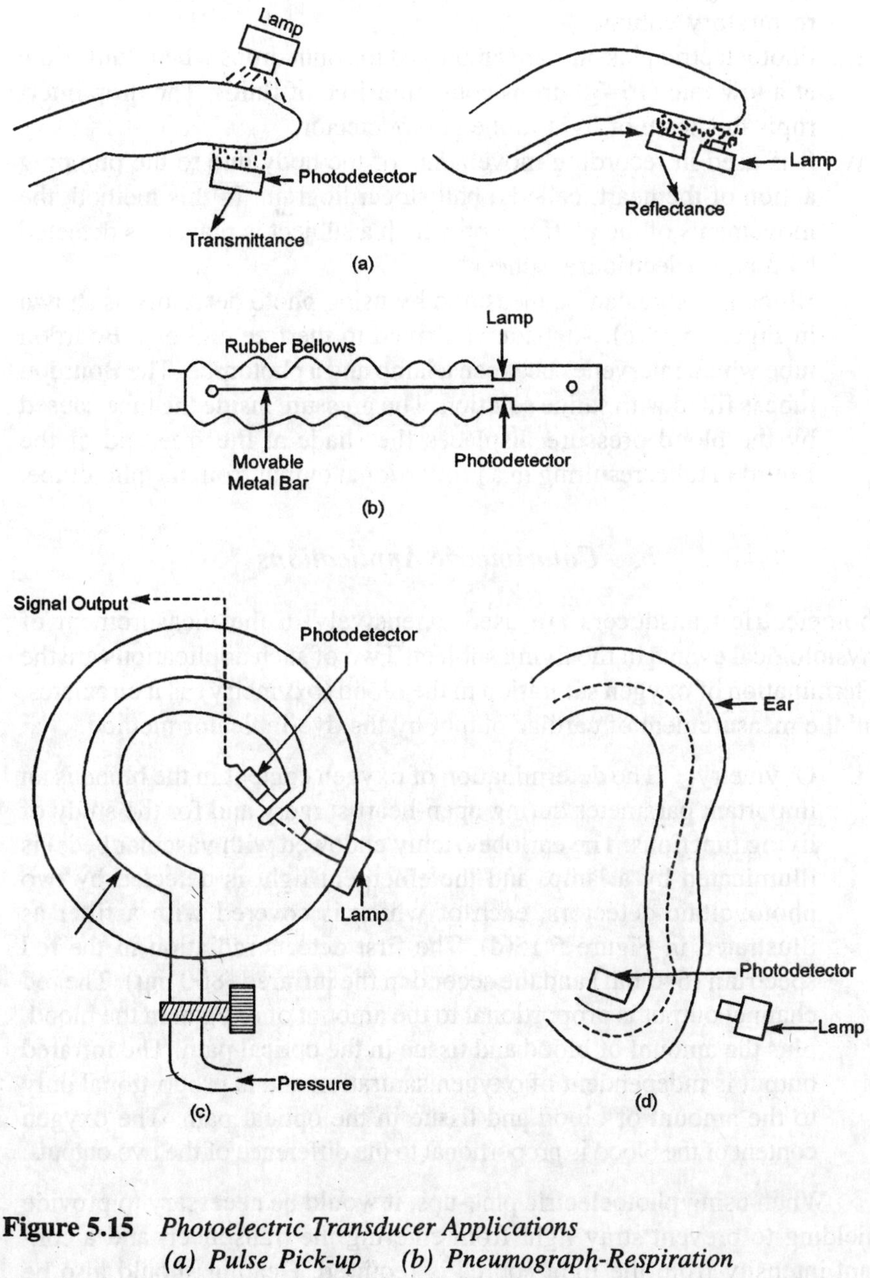

Figure 5.15 *Photoelectric Transducer Applications*
(a) Pulse Pick-up *(b) Pneumograph-Respiration*
(c) Blood Pressure *(d) Ear-lobe Oxymetry*

rubber bellows is wrapped around the chest, and the movable metal bar mounted inside the bellows which causes a shade to alter the amount of light falling on the phototube as the chest expands during breathing. The record, thus obtained, can be calibrated to give the respiratory volume.

iii. Photoelectric pick-ups are employed to count drops when fluids flow at a low rate (15–30 drops constitute 1 cc of fluid). The drop interrupts the beam of light to the photodetector.

iv. It is used in recording movements of the body due to the pumping action of the heart, called a ballistocardiogram. In this method, the movements of the platform on which a subject is placed, is detected by a photoelectric transducer.

v. Blood pressure can be measured by using photo detectors as shown in Figure 5.15(c). A shade is affixed to the free end of a Bourdon tube which intervenes between a lamp and a phototube. The Bourdon tube is filled with saline solution. The pressure inside the tube caused by the blood pressure displaces the shade at the free end of the Bourdon tube, resulting in a proportional output from the phototube.

b. Colorimetric Applications

Photoelectric transducers are used extensively in the measurement of physiological events in the living subject. Two of such applications are the determination of oxygen saturation in the blood (oxymetry) as it circulates, and the measurement of cardiac output by the dye-indicator method.

i. Oxymetry: The determination of oxygen content in the blood is an important parameter during open-heart surgery and for the study of living functions. The earlobe (richly endowed with vascular beds) is illuminated by a lamp, and the emergent light is detected by two photovoltaic detectors, each of which is covered with a filter as illustrated in Figure 5.15(d). The first detects radiation in the red spectrum (640 mμ) and the second in the infrared (800 mμ). The red channel output is proportional to the amount of oxygen in the blood, plus the amount of blood and tissue in the optical path. The infrared output is independent of oxygen saturation and is proportional only to the amount of blood and tissue in the optical path. The oxygen content of the blood is proportional to the difference of the two outputs.

When using photoelectric pick-ups, it would be necessary to provide shielding to prevent stray light from entering the transducer, and a constant intensity from the light source is required. Heating should also be avoided, since heat causes the dilation of blood vessels and, hence, alters the state of the vascular beds under examination. With this device, the patient is isolated from the mains supply and the possibility of earth-leak currents passing through the patient is eliminated.

5.9 Piezoelectric Transducers

Certain crystalline materials, when subjected to mechanical stress, develop surface electric charges. Thus, if a stress is applied to a piece of such material provided with suitable electrodes, charges will appear on the electrodes and hence a voltage will be developed across them as shown in Figure 5.16(a).

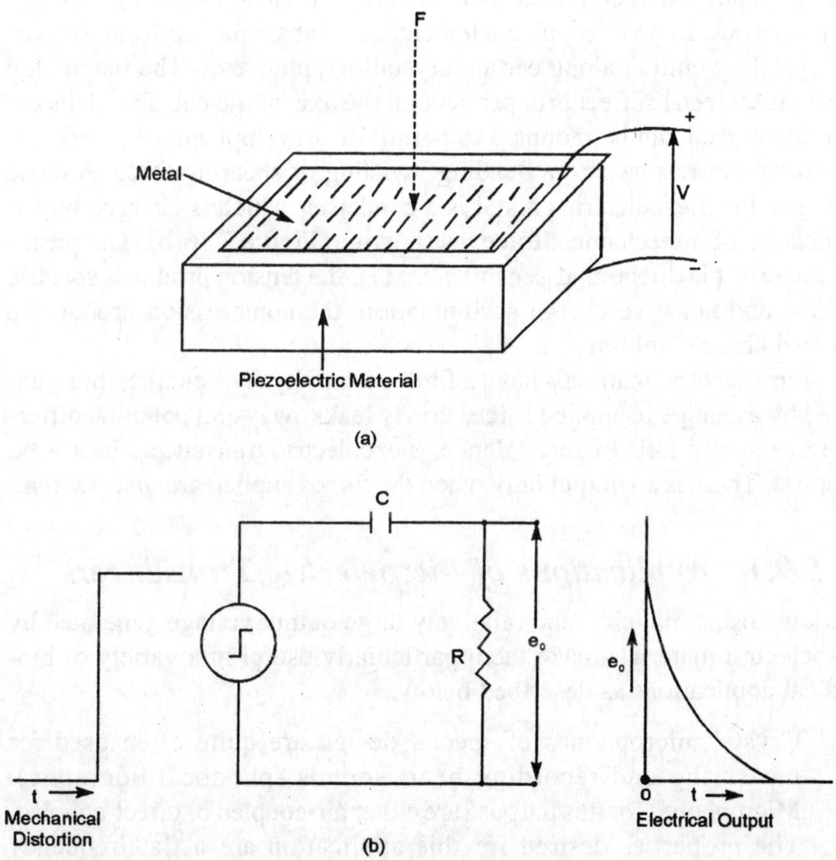

Figure 5.16 *(a) Piezoelectric Transducer Principle*
(b) A Close Analog to the Piezo Crystal

Piezoelectric materials generally possess a linear relation between the change in induced surface charges ΔQ and the change in applied force ΔF, as represented by the relation

$$\Delta Q = D \Delta F \qquad ....(5.10)$$

where D is a constant for the material that expresses its charge sensitivity. A piezoelectric transducer thus behaves like a fixed capacitor carrying a charge proportional to the applied stress. Hence, the piezoelectric transducer is a self-generating or active transducer. The application of an electric field to a piezoelectric crystal distorts it. Thus, the phenomenon is reversible.

Several natural and synthetic materials like quartz, rochelle salt, lithium sulphate and ceramics such as barium titanate possess piezoelectric properties. In practice, piezoelements are slabs removed from the parent crystal by cutting along certain crystallographic axes. The magnitude of the piezoelectric effect is dependent on the axis of the cut. The slabs cut from the crystal can be mounted to permit the development of piezoelectric voltage in response to a bending, twisting or shearing force. A close analog to the piezoelectric crystal is a condenser which is charged by the application of a mechanical force, as shown in Figure 5.16(b). The piezoelectric effect is directional-sensitive, that is, if a tension produces specific positive and negative charge accumulation, the compression produces a reversed charge build-up.

Piezoelectric materials have a finite resistivity. The charge, thus generated by a change in applied force, slowly leaks away and potential difference eventually falls to zero. Hence, piezoelectric transducers lack a DC response. There is an output only when the forces applied are *time-variant.*

5.9.1 Applications of Piezoelectric Transducers

The low cost, simplicity, and relatively large output voltage generated by piezoelectric materials make them particularly useful in a variety of biomedical applications as described below.

1. Crystal microphones of special design are quite often used for measuring and recording heart sounds (phonocardiography). Microphones for this purpose are either air-coupled or direct-coupled. The properties desired for this application are a flat frequency response from 20 to 1000 Hz, and directional sensitivity, so that breathing sounds, room noise and the like are suppressed.
2. Catheter-tip piezoelectric sensors are used for localized pick-up of sounds from within the chambers of the heart (intracardiac

phonocardiography). They are introduced directly into the left heart via a thin-walled needle introduced through a peripheral vein or artery.

3. Piezoelectric crystals are widely used for detecting the sounds (Korotkoff sounds) associated with the occlusive cuff-method for measuring systolic and diastolic blood pressure. The cuff pressure at which these appear and disappear as the pressure is reduced, corresponds to the systolic and diastolic blood pressures, respectively. This is illustrated in Figure 5.17.
4. It is also used for detecting radial pulse.

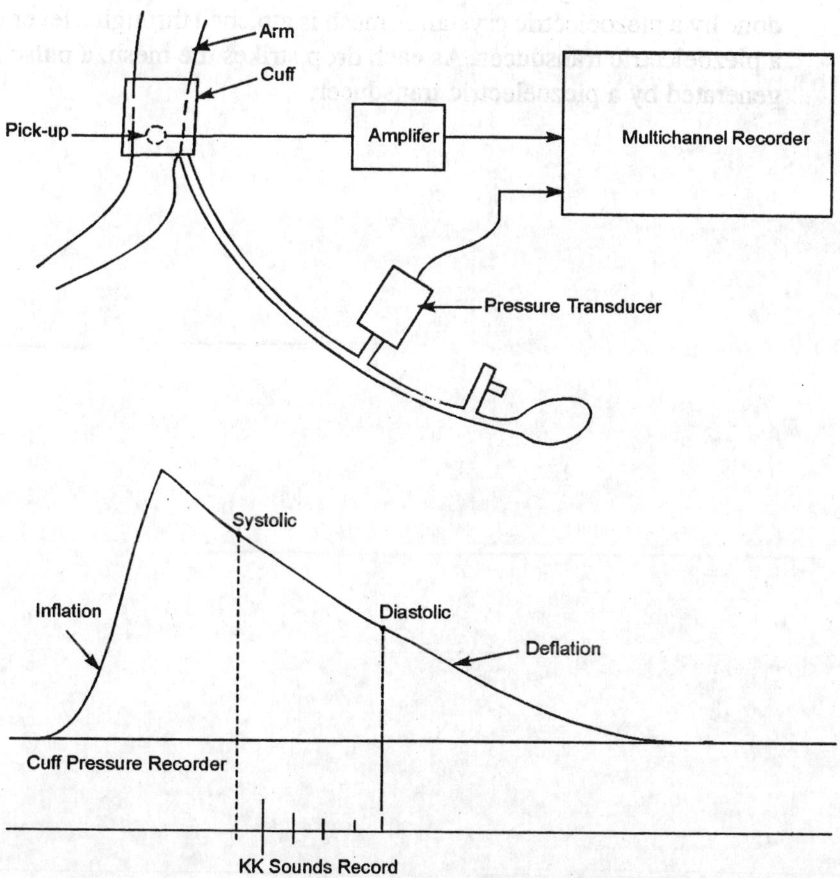

Figure 5.17 *Piezoelectric Transducer for Picking up KK Sounds for Blood Pressure Recording*

5. Piezoelectric crystals are employed in medical ultrasonic devices. Ultrasonic pulses are transmitted into the body structure and an echo from this structure is displayed on a cathode ray oscilloscope, which is used for diagnosis.
6. It is used in ballistocardiography, for detecting the movement of a freely suspended platform on which a patient is made to rest. The platform with the patient moves due to the pumping action of the heart. These tracings are used for diagnosing abnormalities of the heart.
7. They are also used in ultrasonic blood-flow meters using the Doppler effect.
8. When drugs are administered intravenously, it is necessary to know the amount of drug administered. For this purpose, drop-counting is done by a piezoelectric crystal. A mesh is attached through a lever to a piezoelectric transducer. As each drop strikes the mesh, a pulse is generated by a piezoelectric transducer.

Chapter 6

Biomedical Amplifiers

In biomedical measurements, biopotentials are involved that are often at very low levels and have high source impedances. Amplifiers are thus required to boot the biopotentials while maintaining high fidelity so that it may be further processed, recorded or displayed. Amplifiers are used for interfacing transducers that sense body motions, blood pressure, heart sounds, blood flow, *etc.* In this chapter, some of the basic requirements and examples of such amplifiers will be discussed.

6.1 Basic Requirements

Usually such amplifiers are in the form of voltage amplifiers, since they are capable of increasing the voltage of a signal. Some of the requirements are as given below:

1. They must have high input impedance, so that the signal to be measured is minimally loaded, thereby reducing the distortion of the signal. Modern biopotential amplifiers have input impedances of at least 2 MΩ and, for many applications, 10 MΩ is a desirable value.
2. If the amplifier is used to amplify biopotentials picked up by electrodes from the body, the input circuit must also provide protection against any current or potential appearing across the amplifier input terminals. Otherwise, this can cause macro/microshocks in the patient being monitored. The amplifier input must, therefore, have isolation and protection circuitry.
3. The amplifier load is usually an indicating or recording device and, as such, the output circuit of a biopotential amplifier does not present many problems as the input circuit. The output impedance of the amplifier must be low, with respect to the load impedance, to maintain maximum fidelity and the power output of the amplifier must be capable of meeting the load requirement.
4. Biomedical amplifiers must cover the frequency range of biopotentials or the output voltages of transducers. Because of the low level of such signals, it is important to limit the bandwidth of the amplifier

so that it is sufficient to adequately process the signal. Thus, the optimal signal-to-noise ratios can be obtained.

5. Biopotential signals are usually in the milli- or micro-volt range. Biopotential amplifiers must, therefore, have high gains of the order of 1000 or greater.
6. A biopotential differential amplifier is required while using bipolar electrodes to pick up biopotentials. Under such circumstances, the common-mode voltage of bipolar electrodes, with respect to the ground, is much larger than the biosignal to be measured. Such biopotential differential amplifiers must, therefore, have high common-mode rejection ratios (CMRR) to reduce interference due to the common-mode signal.
7. It is necessary, not to analyze the wave forms that are amplified and recorded, but to know their amplitudes. The gain of the amplifier must, therefore, be correctly calibrated.
8. This is provided by a standard signal source that is momentarily connected to the input of the amplifier by such a button, for the calibration of the pen recorder or display unit.

Apart from the above common basic requirements, each specific application calls for additional requirements which are discussed in the chapter to follow.

6.2 Differential Amplifier

In the single-ended amplifiers as illustrated in Figure 6.1, one of the input terminals is connected to common line, *i.e.*, it is connected to the negative of the power supply and to the output terminals. The common line is often referred to as the ground terminal.

In electrophysiology a variety of bioelectric events are recorded simultaneously on a multichannel recorder. For reasons of economy, all channels are operated from the same power supply instead of having a

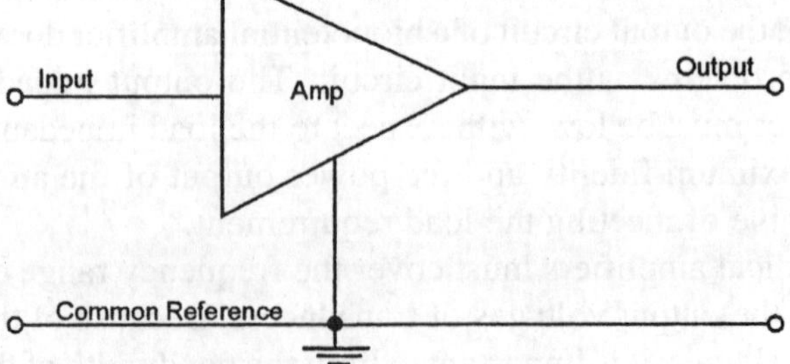

Figure 6.1 *Single-ended Amplifier*

separate source of power supply for each channel. When this is done, all the channels will have one of the input terminals joined together. If multiple electrodes are placed on a subject (say, the human body), all bioelectric voltages picked up are referred to the common electrode. If an event occurs at the common electrode, it appears in all the channels. Thus, with such an arrangement, it is not possible to record the true differences of potentials between pairs of electrodes placed at will on the subject. The need for this freedom of choice was an important reason for the development of the differential amplifier.

Secondly, when making recordings of bioelectric events in the presence of wiring carrying alternating currents, extensive shielding has to be provided. The whole preparation and the amplifier has to be placed in shielded cabinets. This makes it convenient to change the position of the electrode on the subject. Further, in many instances it is not possible to provide adequate shielding. Power lines are ground referred, so it is necessary to connect the shielding to a stable ground connection. When one terminal of the single-sided amplifier is grounded, it often becomes difficult to shield the other from power-line interference. In the differential amplifier, as illustrated in Figure 6.2, neither input terminal is at ground potential. This amplifier provides an output proportional to the difference of potential between the input terminals. Hence, it matters little what the potential of the two are; so long as they are the same, the amplifier will provide zero output.

Thus, in summary, the two main reasons for the use of differential amplifiers in electrophysiology are:

1. Freedom of choice in placing the electrodes to record only the difference of potential that appears between them.
2. Inherent rejection or voltage which appear on both input terminals—such as ground referred interference from power lines.

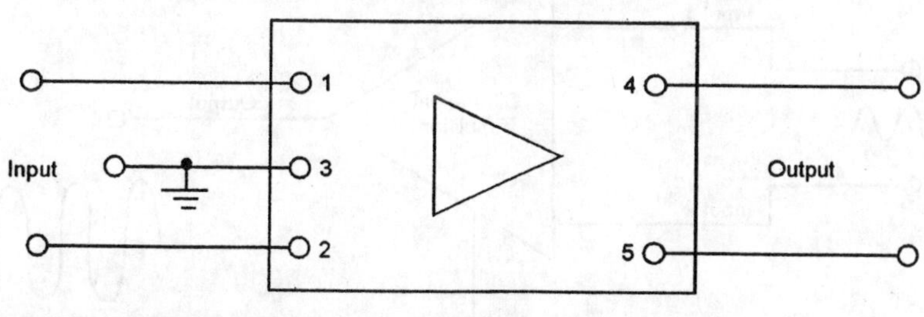

Figure 6.2 *Schematic Illustration of a Differential Amplifier*

If the signal at terminal 1 is 'a' and at 2 is 'b' with terminal 3 (ground) considered as zero reference, the input signal or differential input signal is (a–b).

An example of how the signal may appear in a typical biological application is illustrated in Figure 6.3. The solid line signal 'a' and the

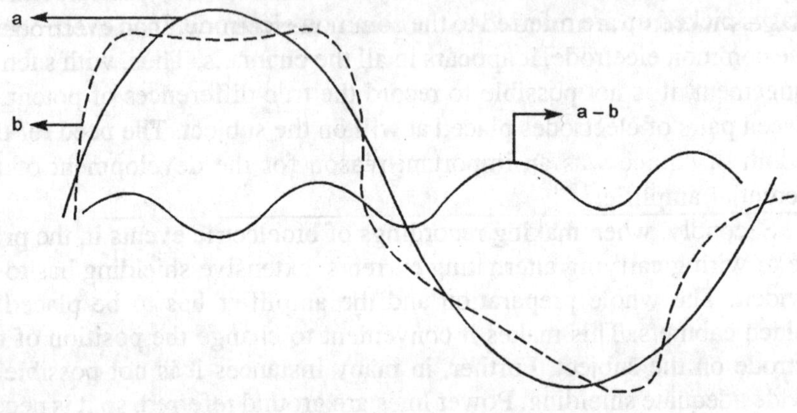

Figure 6.3 *Input Signals of a Typical Biological Application*

dotted line signal 'b' are both large and vary almost, but not quite, together. The difference between them represents the differential input signal and is the low amplitude curve a–b. It is this small difference which one wishes to amplify so that it is unaffected by the much larger (in this case) component, common to the two signals—the common mode or in-phase signal. The degree to which the amplifier is able to reject this common mode signal, determines how well it acts as a true differential amplifier and is, thus, an important feature of the amplifier.

A differential amplifier as shown in Figure 6.4 has 3 inputs, one of which serves as reference for the other two. An input signal can be applied between input 1 and the reference, and another signal can be applied

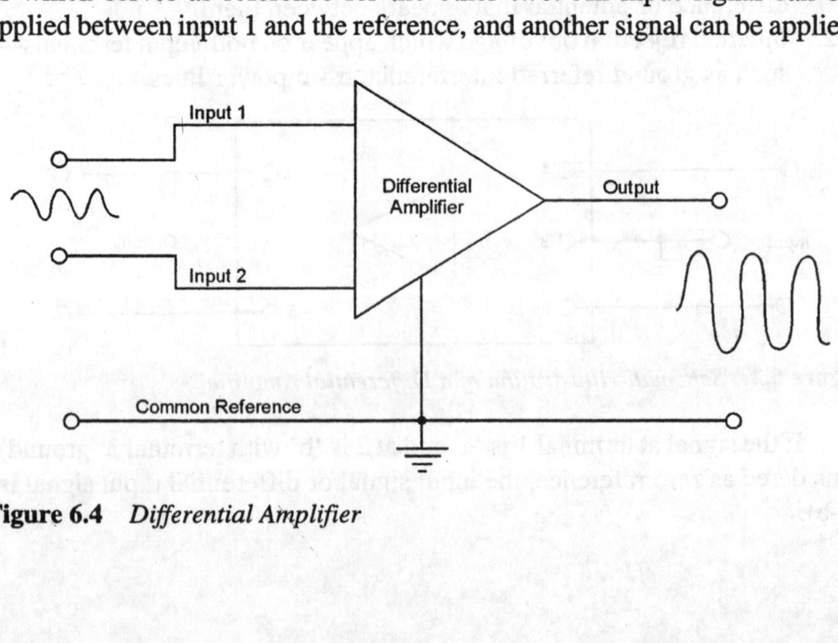

Figure 6.4 *Differential Amplifier*

between input 2 and the reference. The output is an amplified signal of the difference between these two inputs.

To understand the use of the differential amplifier, assume that the voltage which is the difference of potential between two ECG electrodes is to be amplified. In Figure 6.5 electrodes A and B are connected to input 1 and 2 of the differential amplifier, respectively. A third electrode on the (right) leg of the patient serves as a reference electrode and is connected

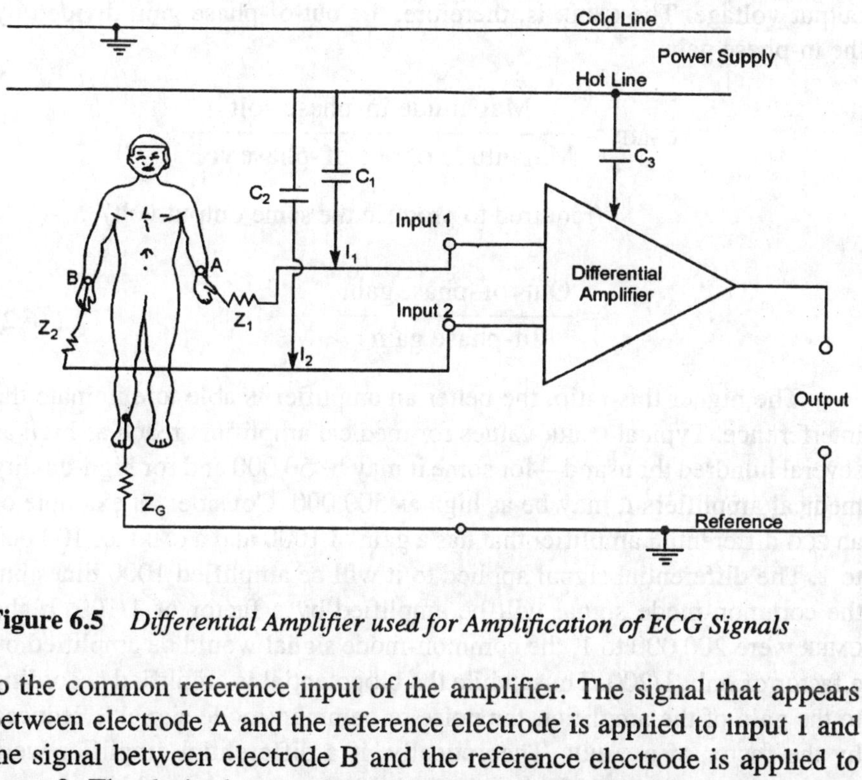

Figure 6.5 *Differential Amplifier used for Amplification of ECG Signals*

to the common reference input of the amplifier. The signal that appears between electrode A and the reference electrode is applied to input 1 and the signal between electrode B and the reference electrode is applied to input 2. The desired ECG voltage, however, is the difference in potential between electrodes A and B which is amplified as shown in Figure 6.4. The interference (unwanted) voltages picked up by each of the two ECG electrodes are almost identical and, when they are applied to the two inputs of the differential amplifier, their difference is amplified. Since this difference in this case will be essentially zero, the interference voltages, even though much larger than the desired signal, are effectively cancelled out and only the desired ECG signal is amplified. If, however, in contrast, if the single-ended amplifier shown in Figure 6.1 was used, both the ECG signal and the interference voltage would be amplified resulting in the desired ECG signal being lost if the interference voltage were much larger.

Since the interference voltage is common to both inputs 1 and 2, it is called a *Common-Mode* input. The attractive feature of the differential amplifier lies not in its voltage gain characteristics, but in its insensitivity to a common-mode input, that is, the voltages which are simultaneously in-phase at the input. The quantity which describes this insensitivity to in-phase or common-mode input, is called the *Common-Mode Rejection Ratio* (CMRR). It is measured as the ratio of the magnitude of the in-phase voltage to the magnitude of out-of-phase voltage required to produce the same output voltage. The CMRR is, therefore, the out-of-phase gain divided by the in-phase gain.

$$\text{CMRR} = \frac{\text{Magnitude in-phase volt}}{\text{Magnitude of out-of-phase volt}}$$

(required to produce the same output volt) ....(6.1)

$$= \frac{\text{Out-of-phase gain}}{\text{In-phase gain}} \qquad \text{....(6.2)}$$

The higher this ratio, the better an amplifier is able to eliminate the interference. Typical CMRR values for medical amplifiers reach as high as several hundred thousand—for some it may be 50 000 and for high-quality medical amplifiers it may be as high as 500 000. Consider an example of an ECG differential amplifier that has a gain of 1000 and a CMRR of 100 000 to 1. The differential signal applied to it will be amplified 1000 times but the common-mode signal will be amplified by a factor of 1/100. If the CMRR were 200 000 to 1, the common-mode signal would be amplified by a factor of only 1/200. Thus, while the biopotential is amplified according to the gain of the amplifier, the noise or common-mode signal is reduced by the factor of a CMRR. Theoretically, if a differential amplifier were perfect, the CMRR would be infinite and the noise signal would be zero.

The 50 Hz power supply system is the chief source of electrical interference which is coupled into the amplifier through the capacitance existing between the hot side of the power line, the patient, the lead wires and the electrocardiograph machines as represented in Figure 6.5. The current through the capacitance C_3, coupling the hot side of the power line and the electrocardiograph, itself flows to the ground and does not cause interference. C_1 represents the capacitance between the power line and one of the leads. Current I_1 does not flow into the electrocardiograph because of its high input impedance, but rather through the skin-electrode resistances Z_1 and Z_G to the ground. Similarly, I_2 flows through Z_2 and Z_G to the ground.

The voltage amplified is that appearing between input 1 and 2, that is $V_1 - V_2$.

$$V_1 - V_2 = I_1Z_1 - I_2Z_2 \qquad ....(6.3)$$

If the two lead wires run near each other, $I_1 \simeq I_2$, then,

$$V_1 - V_2 = I_2(Z_1 - Z_2) \qquad ....(6.4)$$

If the cables are 9 m long, it has been found that $I_1 \simeq 6$ nA. Skin-electrode impedances may differ by as **much as** 20 KΩ. Therefore,

$$V_1 - V_2 = (6\text{ nA})(20\text{ K}\Omega) = 120\ \mu\text{V} \qquad ....(6.5)$$

120 μV is an objectionable level of interference. This can be reduced by shielding the lead wires and grounding each shield at the electrocardiograph. The skin-electrode impedances, if lowered, will also help in minimizing the interference.

6.3 Carrier Amplifier

Many transducers based on changes in inductances or capacitance, require sinusodial AC excitation for operation. Also, resistive devices, such as strain gauges, may often be utilized more effectively if operated with AC. To measure steady or slowly changing events (as in biomedical research), it is desirable or necessary to use transducers excited by AC. Devices known as carrier amplifiers permit the amplification of slowly changing (low-frequency) events and those containing static AC components by means of stable, high gain AC amplifiers. Carrier amplifiers possess the advantage of operating over a very narrow band of frequencies. By so limiting the bandwidth requirements of the amplifier, noise either internal or external to the device which falls outside the pass band of the amplifier does not appear in the output. The ratio of signal to noise is increased accordingly.

The principle underlying the carrier amplifier consists of causing the low frequency event to modulate a high frequency (carrier) signal used to energize the transducer. After amplification, the modulated carrier is subjected to an appropriate detection process for recovery of the desired low frequency modulating signal. The principle of the carrier amplifier as employed with strain gauge transducers arranged in a Wheatstone bridge configuration, is shown in Figure 6.6.

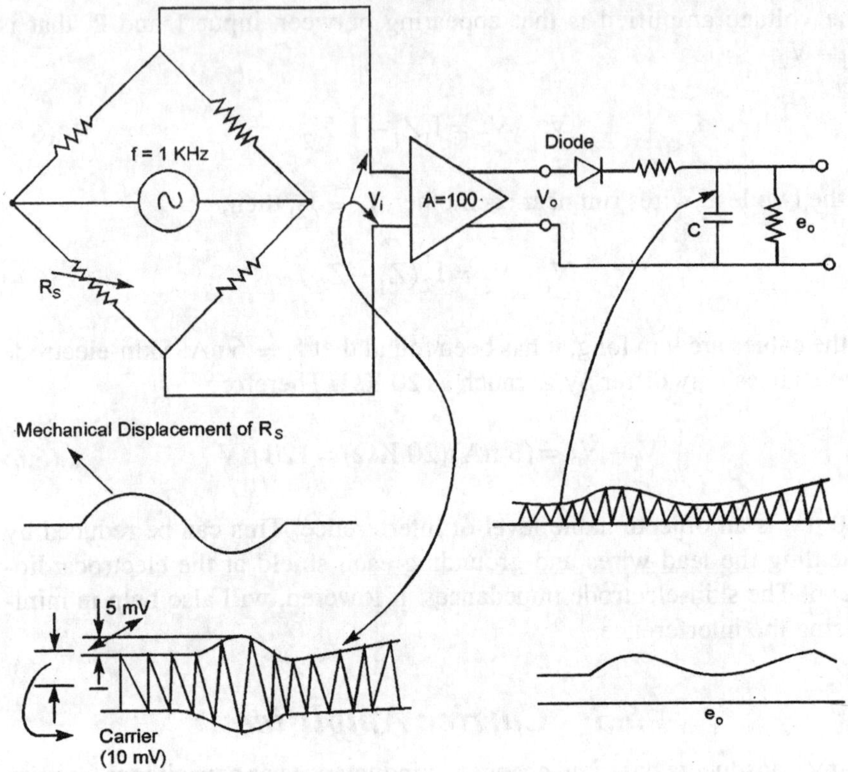

Figure 6.6 *Carrier Amplifier as Employed with Strain Gauge Transducers in a Wheatstone Bridge Configuration*

The bridge may be made of one or more strain gauge elements suitably arranged to sense strain. As an example, consider that the strain gauge bridge is excited from a 100 Hz sinusoidal source and is not balanced to null but is rather slightly offset to provide a 1000 Hz voltage of 10 mV peak, when the strain gauge R_s is in the arbitrarily designated no strain or "zero" position. Assume also that the relationship between the strain gauge movement and the corresponding change in resistance is known and is linear. If the strain gauge is displaced mechanically in a sinusoidal manner at a low frequency rate (*e.g.*, 1 Hz), the change in resistance will be sinusoidal and, if the input impedance of the amplifier is very high with respect to the resistance of the bridge, the amplifier will not load the bridge (draw current). The amplitude of the 1000 Hz output voltage from the bridge will be modulated sinusoidally by the resistance change. Assume that the modulation changes the bridge output from 10 mV for zero displacement to 15 mV for maximum displacement in the opposite direction, as shown in Figure 6.6. If the gain of the amplifier is 100, the variation in the voltage

V_2 appearing at the amplifier output terminals, are similar in form but 100 times larger in amplitude. A diode is used at the output of the amplifier to convert the voltage V_0 into unidirectional pulsations smoothed by the RC circuit which, if it possesses a suitable time constant, effectively short circuits the carrier and higher-frequency components and leaves only the steady DC component and the slowly alternating (1 Hz) components.

In order to ensure a faithful reproduction of the input signal, the carrier frequency should be at least 10 times that of the highest significant frequency component contained in the signal, that is, many carrier cycles must occur in regions where the input signal changes rapidly.

Thus, with carrier amplifiers, the characteristics of DC amplification may be obtained with DC amplifiers.

6.4 Chopper Amplifier

DC amplifiers exhibit a certain amount of 'drift'—the output will shift continually even though the input is zero or constant. In sensitive amplifiers, this drift may be quite troublesome. The measurement of low-level DC signals or slowly changing voltages is commonly encountered in biomedical studies.

A solution to the problem of drift and stability in DC amplifiers is the use of the chopper amplifier, the schematic diagram of which is shown in Figure 6.7.

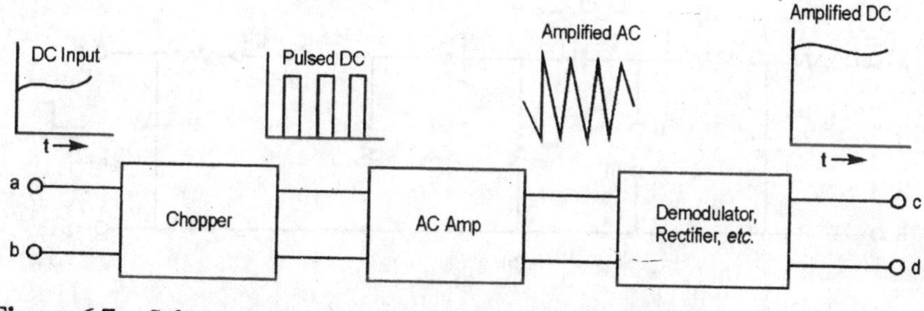

Figure 6.7 *Schematic Operation of the Chopper Amplifier*

The sequence of operation amounts to:

1. a method for converting the DC signal to an AC signal (chopper),
2. an AC amplifier for boosting this AC signal to the desired level,
3. a method for converting the AC signal back to DC (demodulator, rectifier, *etc.*).

The basic operation of the chopper amplifier may be understood from Figure 6.8(a). The chopper, S_1, alternatively connects the input terminal C of the AC amplifier to terminal B which is connected to the ground. When connected to B, the amplifier input terminals are short-circuited and the input voltage is zero. The input to the amplifier consists, therefore, of an AC voltage varying from zero to the value of the input voltage existing when the signal is connected (S_1 open). By this process, a steady DC or slowly varying signal is 'chopped' into a 'train of pulses', having a frequency equal to the rate of the chopper. After amplification, the chopped signal is rectified with a diode. The rectified signal is then filtered and the DC and slowly varying signal components are recovered.

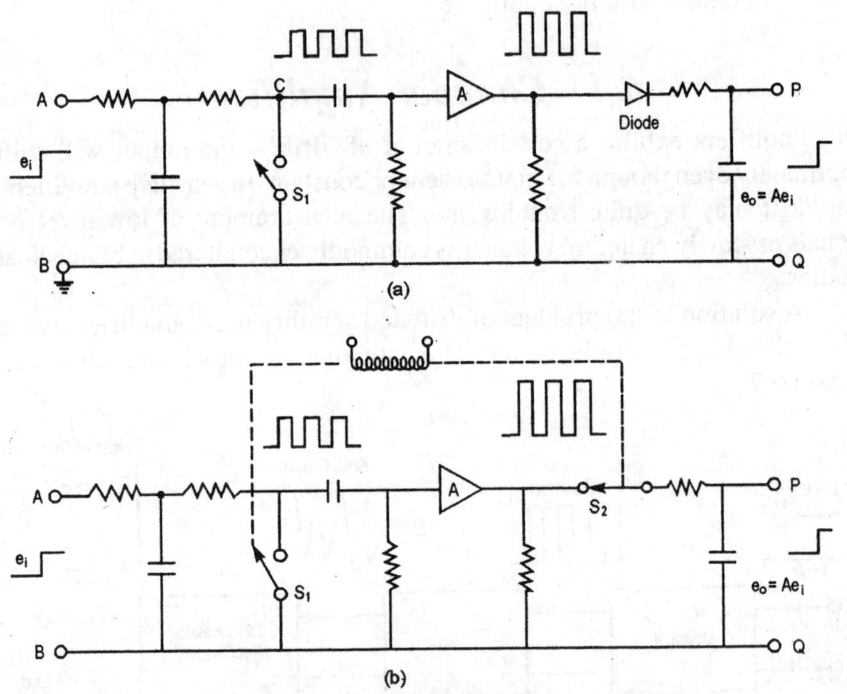

Figure 6.8 *(a) Mechanical Chopper and Diode Rectifier*
(b) Synchronous Mechanical Choppers

Instead of rectifying the output of the amplifier with a diode, a second chopper contact operating synchronously with the first may be used, as shown in Figure 6.8(b).

The rapidity of the movement of mechanical choppers are limited due to their inertia. They are available in 60 and 400 Hz units. These devices have a finite life.

Chopping by the semiconductor switches is now used, with a subsequent increase in chopping rate. Such devices are not subject to mechanical wear and have an extremely long life. Semiconductor choppers may be in the form of conventional transistors which are switched from a conducting to non-conducting state (*i.e.*, operated in the switching mode). An example is shown in Figure 6.9(a) using NPN transistors. The circuit will conduct when 1 is positive and not conduct when 1 is negative. The transistor chopper has the advantage that it will work at many thousands of cycles per second, and will operate with a practically unchanged performance for many thousands of hours. Mechanical choppers may require readjustments after several thousands of hours of use.

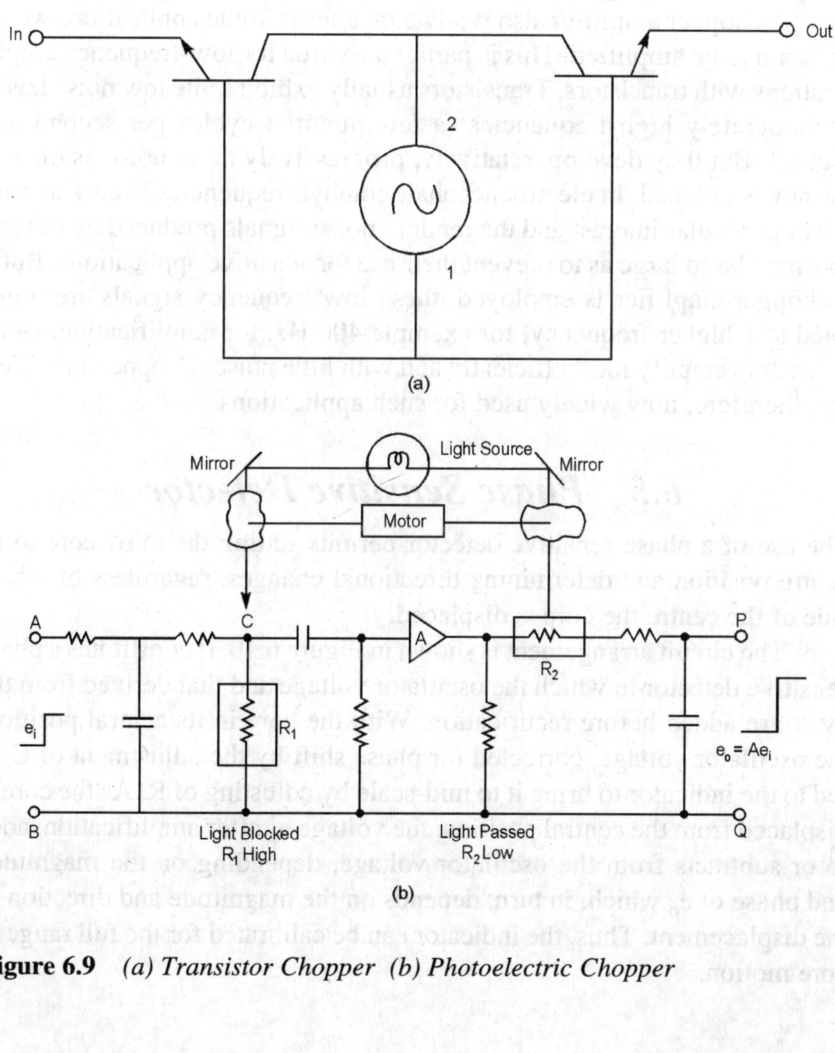

Figure 6.9 *(a) Transistor Chopper (b) Photoelectric Chopper*

A type of chopper which is excellent for modulating low-level signals is the photoelectric chopper using photoresistors. The photoresistor has a very high resistance in the dark, but a resistance ranging from a hundred to thousands of ohms when illuminated. Thus, two photoresistors may be used in place of two contacts of a mechanical chopper, each unit being alternately illuminated from a source of radiant energy by the use of a shutter arrangement, as shown in Figure 6.9(b). The response time of the chopper amplifier is governed by the chopping or sampling rate. If the signal is changing slowly, and there occurs a sharp notch, it is necessary to use a high chopping rate, that is, many cycles of the chopper are required to sample the notch. The rule commonly applied is that the sampling rate should be at least ten times greater than the highest sinusoidal frequency component of the signal.

A chopper amplifier also is advantageous in some applications where DC is not to be amplified. This is particularly true for low-frequency amplifications with transistors. Transistors usually exhibit quite low noise levels at moderately high frequencies (a few hundred cycles per second and higher). But they develop, relatively, progressively more noise as the frequency is reduced. In electroencephalography, frequencies from 1 to 5 Hz are of particular interest and the random noise signals produced by transistors may be so large as to prevent their use for sensitive applications. But if a chopper amplifier is employed, these low frequency signals are translated to a higher frequency, for example 400 Hz, for amplification. Here, transistors amplify most efficiently and with little noise. Chopper amplifiers are, therefore, now widely used for such applications.

6.5 Phase Sensitive Detector

The use of a phase sensitive detector permits setting the LVDT core to its centre position and determining directional changes, regardless of which side of the centre the core is displaced.

The circuit arrangement is shown in Figure 6.10. It constitutes a phase sensitive detector in which the oscillator voltage and that derived from the LVDT are added before rectification. With the core in its central position, the oscillator voltage, corrected for phase shift by the adjustment of C, is fed to the indicator to bring it to mid-scale by adjusting of R. As the core is displaced from the central position, the voltage e_o after amplification, adds to or subtracts from the oscillator voltage, depending on the magnitude and phase of e_o which, in turn, depends on the magnitude and direction of the displacement. Thus, the indicator can be calibrated for the full range of core motion.

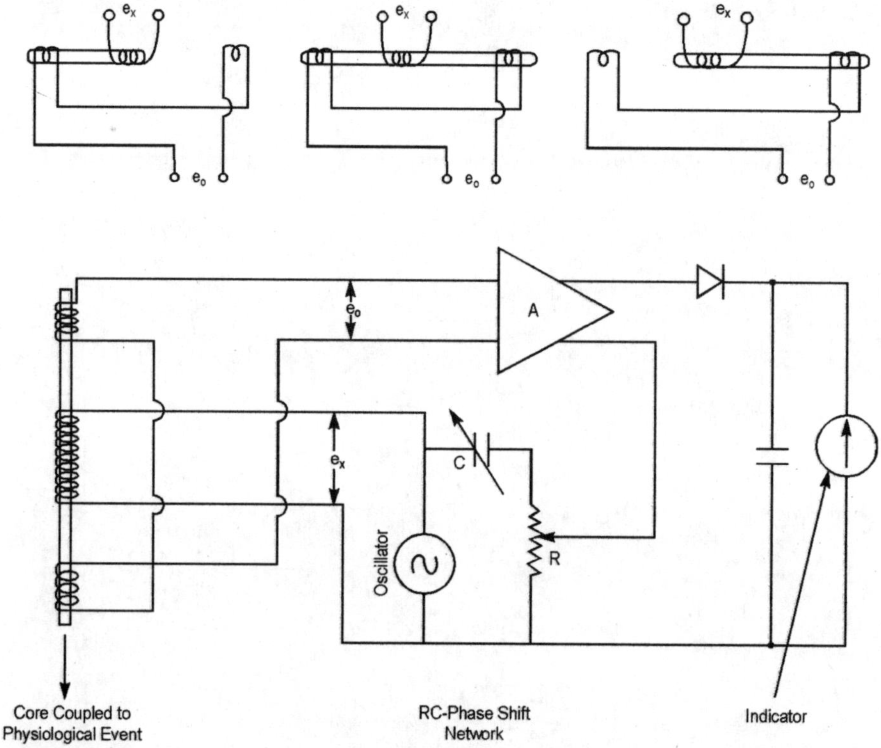

Figure 6.10 *Phase Sensitive Detector for LVDT*

Chapter 7

Principles of Recorders for Recording Bioelectric Events

It has been discussed in Chapters 4 and 5 how a physiological variable is transduced into a small electrical signal and amplified. It is necessary to see and analyze the biosignal thus obtained. For this purpose, the output of the amplifier has to be connected to a suitable read-out, writing or oscillograph device. A continuous visual record of physiological data versus time may either be obtained with a graphic recorder or with an oscilloscope and a continuous motion camera. Continuous motion CRT photography, while offering an almost unlimited bandwidth and extremely fast recording rates, does not provide an immediate record of the recorded data. A graphic recorder, while inherently possessing limited high-frequency response characteristics, does provide an immediate record of the recorded data for observation. These recorders are often referred to as oscillographic recorders, stripchart recorders or chart recorders. Graphic recorders utilize some form of stylus to traverse a strip of chart-paper while the chart-paper is in motion, leaving a visible tracing which is a record of the time variations of the input voltage or current.

Most graphic recorders consist of the following five basic components:

1. An electromechanical device to convert an electrical input signal to a mechanical movement,
2. A stylus arm to transmit the mechanical movement from the electromechanical device to the stylus,
3. A stylus to leave a written record on a chart-paper as the stylus moves across the chart-paper,
4. A chart-paper assembly consisting of a chart-paper supply roll, a chart-paper writing table and a chart-paper take-up roll, and
5. A paper-drive mechanism to move the chart-paper across the writing table from the supply roll to the take-up roll at a constant speed.

Normally, each doctor performs his/her diagnosis based on the output from the recorder. If the recording is not faithful or the fidelity is very poor, then there is an improper diagnosis leading to improper therapy and so on. Thus, the recorders play a vital role to interpret the nature of the biosignal source during its working.

7.1 Characteristics of the Recording System

For a faithful reproduction of the input signal, the following three basic conditions must be satisfied by the individual parts of the system.

1. A recorder is said to be *linear* if the pen deflection is proportional to the amplitude of the input signal. The range over which the system is linear is called the dynamic range. Beyond the dynamic range, the non-linear distortion occurs. Non-linearity can occur separately in the amplifier and the writer. Electrode potentials at the input may be the source of amplifier non-linearity.
2. The *frequency* response of the recorder is the frequency characteristics of the system, such as variation of the sensitivity for different frequency components of the biosignal. The *sensitivity* of the recording system is the magnitude of the input voltage required to produce a standard deflection in the recorded trace. Typically, it is about 5 UV per mm in the case of an EEG recording. The sensitivity is composed of the gain of the amplification system and the sensitivity of the writer in the recorder. There are provisions in the recorder to change the sensitivity. A recorder is said to have a good frequency response when the sensitivity of the system is constant for all the frequencies present in the signal, with different amplitudes, and amplification should be done without any distortion. Normally, there is a band of frequencies over which the sensitivity is constant. The frequencies at which the sensitivity has dropped to 70.7% of the maximum are called the turnover frequencies or half-power points. The interval between turnover frequencies is the bandwidth of the system.
3. *The phase response* of a recorder is measured by means of the time delay between the input and output. A complex waveform will be distorted if the time displacements introduced are different for different frequencies. The noise level increases with the bandwidth of the system. It is, therefore, important to restrict the bandwidth to that required for a faithful reproduction of the signal. A low noise level is maintained by special circuits and use of quality electronic components.

7.2 The Analog Graphic Recorders

There are three main types of analog recorders: (a) the heated stylus, (b) the optical-light, and (c) the ink-jet. All of these recorders use a galvanometer to convert an electrical signal to a mechanical motion.

The stylus galvanometers used in recorders are of two types: the moving coil recorder and the moving iron recorder.

7.2.1 Moving Coil Recorder

This is illustrated in Figure 7.1. It works on the principle of the d'Arsonval-type direct writing oscillograph, in which the coil C is pivoted to turn between the poles of a permanent magnet. The output of the amplifier is connected to this coil and the passage of current through this coil (sent through a suspending wire situated in the magnetic field) produces a force tending to turn it against the restraining force of spring, R, which restores the coil to its central position when the current ceases to flow The pen arm, P, rotates over the recording chart due to the motion of the coil. The iron core, X, improves the magnetic coupling with the coil. The frequency response of this system is limited to about 70 Hz.

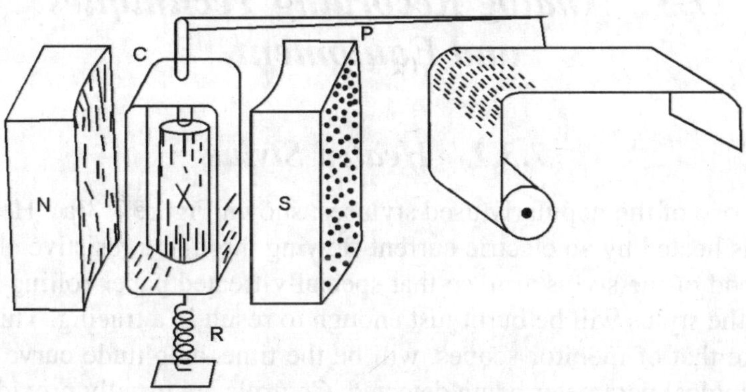

Figure 7.1 *The Principle of a Moving Coil Recorder*

7.2.2 Moving Iron Recorder

In this type, illustrated in Figure 7.2, the coil C is fixed and the iron armature A is attached to the pen shaft P. The iron armature A is attracted to the iron core with a force depending on the strength of the current through the coil C. The frequency response of this system is similarly limited as in the moving coil recorder. In both the systems described above, either a heated stylus or ink-pen arm is necessary to write the signal on a suitable paper.

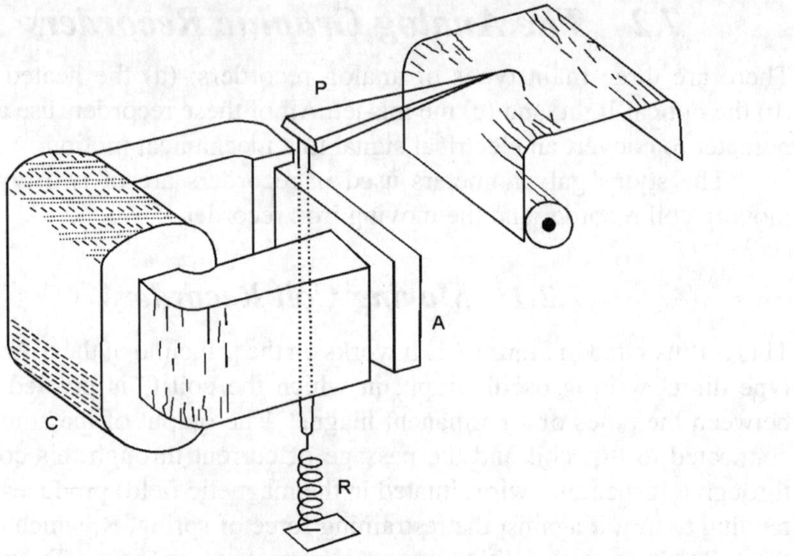

Figure 7.2 *The Principle of a Moving Iron Recorder*

7.3 Analog Recording Techniques and Equipment

7.3.1 Heated Stylus

This is one of the popularly used styluses, shown Figure 7.3(a). Here, the stylus is heated by an electric current flowing through a resistive element at the end of the stylus arm, so that specially-treated paper rolling by the end of the stylus will be burnt just enough to result in a tracing. This tracing, like that of monitor scopes, will be the time–amplitude curve of the physiological parameter being detected. Controls are usually provided for adjusting the heat of the stylus, the amplitude of the swing of the stylus, and the speed of the paper. While this type of stylus recorder requires little maintenance, the specially-treated paper is somewhat expensive, making this technique unsuited to applications requiring long-term recording.

7.3.2 The Optical-light

The optical-light/galvanometer technique, Figure 7.3(b), works on the principle of a rotating mirror which reflects light onto a special photographic paper, thereby producing the time–amplitude curve of the physiological

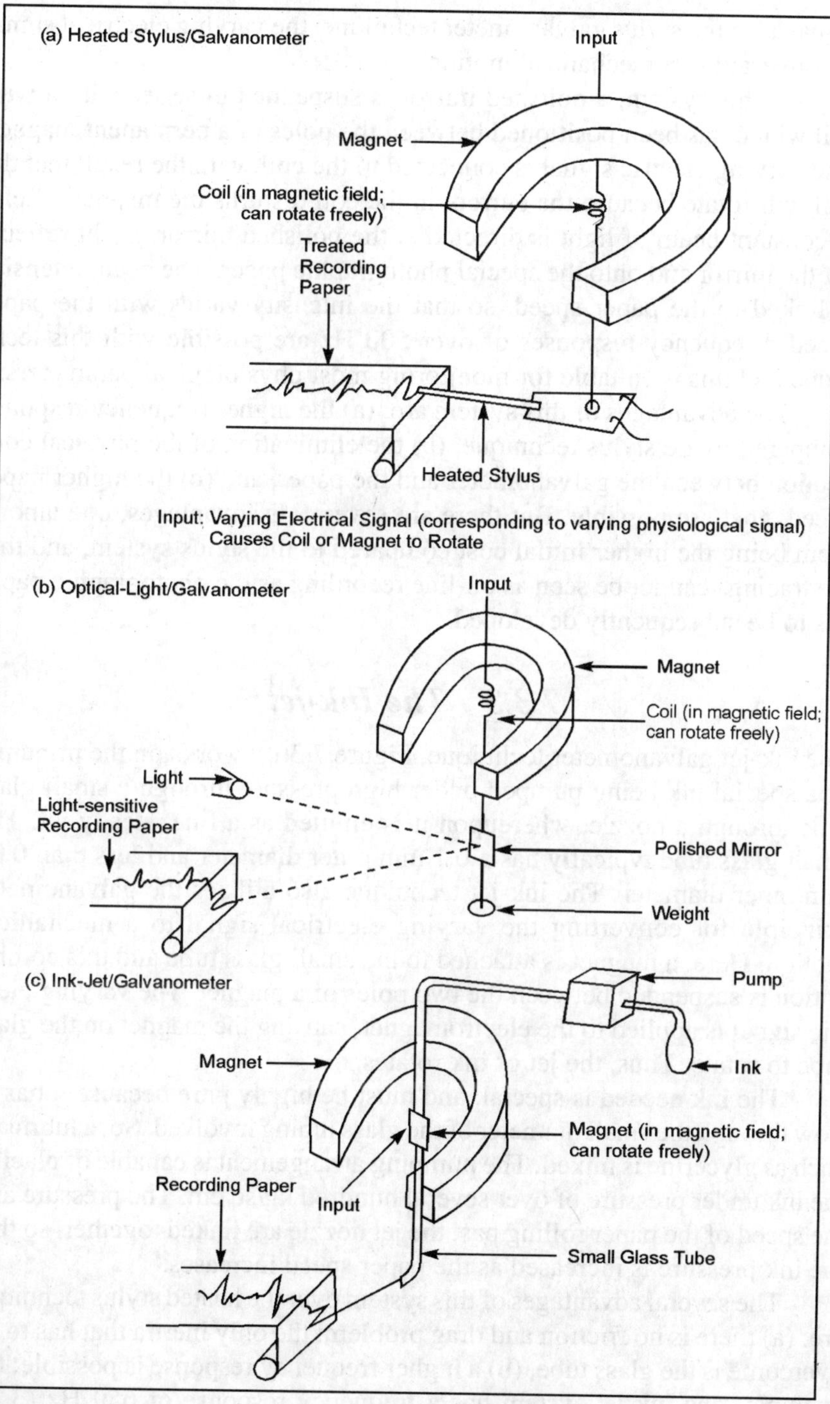

Figure 7.3 *Analog Recording Techniques and Equipment*

signal as in the stylus galvanometer technique, the varying electrical signal is converted to a mechanical motion.

In this system, a polished mirror is suspended in series with a wire coil which has been positioned between the poles of a permanent magnet. The varying electric signal is connected to the coil, with the result that the coil will rotate because the current in the coil disturbs the magnetic field. A constant beam of light is directed at the polished mirror. Light reflects off the mirror and onto the special photographic paper. The beam intensity is linked to the paper speed, so that the intensity varies with the paper speed. Frequency responses of over 500 Hz are possible with this technique and this is suitable for monitoring most physiological parameters.

The advantages of this system are: (a) the higher frequency response compared to the stylus technique, (b) the elimination of the physical connection between the galvanometer and the paper, and (c) the higher paper speeds that are possible. But there are several disadvantages, one among them being the higher initial cost compared to the stylus system, and that the tracings cannot be seen as on-line recording as the photographic paper has to be subsequently developed.

7.3.3 The Ink-jet

The ink-jet galvanometer technique, Figure 7.3(c), works on the principle of a special ink being pumped under high pressure through a small glass tube through a nozzle, whereupon it is emitted as a fine, steady jet. The small glass tube typically has a 0.1 mm outer diameter and less than 0.01 mm inner diameter. The ink-jet technique also utilizes the galvanometer principle for converting the varying electrical signal to a mechanical motion. Here, a magnet is attached to the small glass tube and this combination is suspended between the two poles of a magnet. The varying electric signal is applied to the electromagnet, causing the magnet on the glass tube to rotate. Thus, the jet of ink rotates.

The ink needed is special, and must be highly pure because it has to flow through the small diameter of the glass tubing involved. So, a lubricant such as glycerine is mixed. The pumping arrangement is capable of placing the ink under pressure of over several hundred kg/sq cm. The pressure and the speed of the paper rolling past the jet nozzle are linked together, so that the ink pressure is increased as the paper speed increases.

The several advantages of this system over the heated stylus technique are: (a) there is no friction and drag problem; the only inertia that has to be overcome is the glass tube, (b) a higher frequency response is possible; for example, one ink-jet system has a frequency response of 650 Hz (3 db down) compared to most heated stylus responses of 100 Hz (3 db down), and (c) the recording paper does not have to be specially treated. Some disadvantages include: (a) the ink-jet system is more intricate and sensitive, and (b) the initial cost is higher than the heated stylus.

Some modern electrocardiographs and other recording machines are based upon a recording principle which does not have moving parts in the deflection system. That is, the galvanometers, recording pens and rotating optical mirrors are replaced by stationary, switched systems. These systems operate as follows:

- In one form, the moving paper is a heat-sensitive material with better heat sensitivity than the standard ECG recording paper. Along the Y-axis of the recording, a series of pins are placed at small interval distances such that the entire width of the paper is covered. The application of a pulse of electrical voltage to any one of these pins leads to a mark being produced on the paper at the point where the pin touches the paper. Hence, activating specific pins can produce marks at any point across the width of the paper along the Y-axis. Each of these pins correlates with a particular voltage level.
- The input signal is amplified and the threshold gated with respect to the voltage level of each pin. Pin-activation selection is performed by means of an exclusive OR logic. For any particular voltage, the pin having a correspondence to a voltage just higher than the input voltage level is selected and the pin is activated to produce a mark on the paper. Thus, as the paper moves, the waveform of the input voltage is recorded by the activation of the pin.
- Since there are no moving parts in the marking system, the frequency response can be high. The threshold detection operation and gating take very little time. The actual marking process is also fast. A flat frequency response over 0 to 200 Hz, or more, can readily be obtained. The system, however, has a limitation in that the spatial resolution is limited by the number of pins that can be accomodated across the width of the paper. But the number is sufficiently high, so that the plot appears to be smooth on direct visual observation and the steps can only be detected on optical magnification—which is generally not required.

Chapter 8

Electrocardiography

Electrocardiography is the art of analyzing the electrical activity of the heart by measuring the potential at the surface of the body. This is achieved by applying electrodes at certain positions on the body and recording the potentials generated between various combinations of these electrodes with an amplifier and CRT display or a strip-chart recorder.

8.1 Electrocardiographic Planes

The heart can be regarded as an electrical generator enclosed in a volume conductor, the torso. As this electrical generator is completely enclosed by the torso, a direct measurement of the generator output voltage is impossible without resorting to surgery. The electrocardiographer measures the potential existing between various points on the surface of the volume conductor and uses the information obtained to determine the clinical condition of the heart. The art of electrocardiography is simplified somewhat by considering that at one time the cardiac potential is projected along axes existing on each of three reference planes: the frontal plane, the transverse plane and the sagittal plane as shown in Figure 8.1. This projection of the

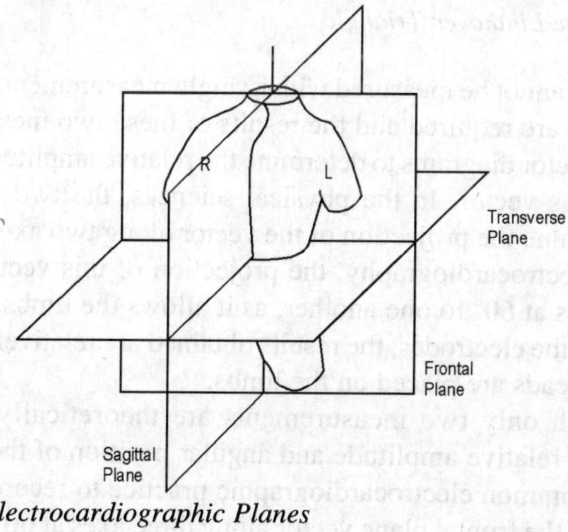

Figure 8.1 *Electrocardiographic Planes*

cardiac potential is referred to as the electrocardiogram (ECG). Various techniques are used to measure the projection of the cardiac potential along the axes existing on each of the three planes.

8.2 Frontal Plane ECG Measurements

The electrical potential generated within the heart, as projected along the axes existing on the frontal plane of the body, is known as the *frontal plane cardiac vector,* as shown in Figure 8.2. As with vector measurements in the physical sciences, the relative amplitude and angular position of this vector

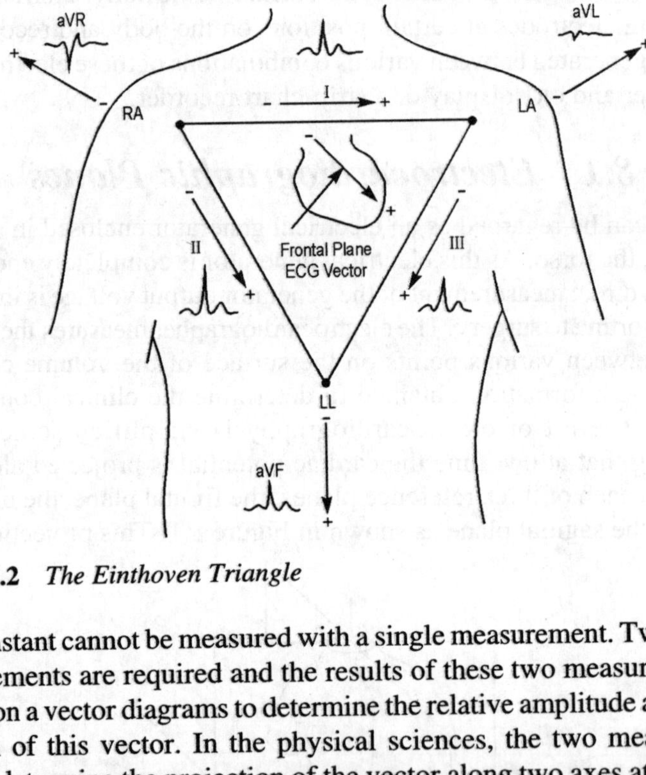

Figure 8.2 *The Einthoven Triangle*

at any instant cannot be measured with a single measurement. Two separate measurements are required and the results of these two measurements are plotted on a vector diagrams to determine the relative amplitude and angular position of this vector. In the physical sciences, the two measurements usually determine the projection of the vector along two axes at 90° to one another. In electrocardiography, the projection of this vector is recorded along two axes at 60° to one another, as it allows the limbs to be used for attachment of the electrodes; the results obtained are relatively independent of where the leads are placed on the limbs.

Although only two measurements are theoretically necessary to determine the relative amplitude and angular position of the frontal plane vector, it is common electrocardiographic practice to record at least three projections of the frontal plane vector along three axes at 60° to each other.

The triangle formed by these three axes is known as the *Einthoven triangle,* as shown in Figure 8.2. William Einthoven, a Dutch physiologist, pioneered many of the electrocardiographic techniques in use today and, during the first quarter of this century, he developed this vector approach to electrocardiography. Einthoven's law states that the vector sum of the projections of the frontal plane cardiac vector at any instant onto the three axes of the Einthoven triangle will be zero. This is a well-known physical law; however, Einthoven realized that it did, in fact, also apply to cardiology. Although the above applies to the cardiographic vector at any instant, cardiologists are usually only interested in the cardiac vector at the peak of the 'R wave'. Thus, when the term "cardiac vector" is used, it implies: "Cardiac vector at the peak of the R-wave".

More recent advances in electrocardiography have pointed out the well-known, but previously ignored, fact that the human torso is neither homogenous nor triangular and that this leads to a distortion of the electric field. Thus, the angle of the vector deduced from Einthoven's triangle would be an error. Burger attempted to allow for the inhomogeneities of the torso by the introduction of the distorted triangle, shown in Figure 8.3, which compensates for the effect of the lungs and the spine.

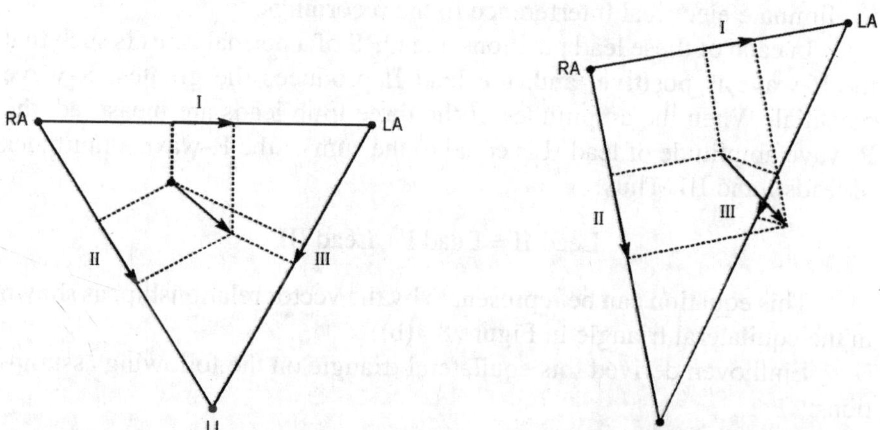

Figure 8.3 ***Einthoven's Triangle Assuming a Homogenous Torso, and Burger's Triangle with the Effects of the Lungs and Spine***

Although Burger's triangle is perhaps a more accurate representation of the frontal-plane cardiac vector, it is not widely used and cardiologists still prefer Einthoven's approach. They realize that Einthoven's vectors are only an approximation, but use them largely for convenience. The clinical interpretation of ECGs is quite empirical in practice, being done by reference to the enormous number of records which have been correlated with known cardiac disorders, usually at the time of autopsy.

8.3 Lead Systems for Recording ECG

8.3.1 Standard Bipolar Leads

One of the most important method is that discovered by Einthoven, who selected the bipolar standard leads (I, II and III) to record the ECG in the frontal plane. Electrodes are applied to the left arm, right arm and left leg and the bipolar leads represent a potential difference (P.D.) between two selected sites, as follows and as shown in Figure 8.4(a).

LEAD – I : P.D. between left arm and right arm (LA - RA)
LEAD – II : P.D. between left leg and right arm (LL - RA)
LEAD – III : P.D. between left leg and left arm (LL - LA)

An electrode is also placed on the right leg and is earthed through the electrocardiograph machine for the electrical protection of the patient and to eliminate electrical interference in the recordings.

In each of these lead positions, the QRS of a normal heart is such that the R-wave is positive, and, the lead II produces the greatest R-wave potential. When the amplitudes of the three limb leads are measured, the R-wave amplitude of lead II is equal to the sum of the R-wave amplitudes of leads I and III. Thus,

$$\text{Lead II} = \text{Lead I} + \text{Lead III}$$

This equation can be represented by the vector relationship, as shown in the equilateral triangle in Figure 8.4(b).

Einthoven derived this equilateral triangle on the following assumptions:

1. Each instantaneous vector is the result of an electrical potential which acts as a dipole.
2. The centre of the electrical activity of the heart is located in the anatomic centre of the heart.
3. The LA, RA and LL are all equidistant from the centre of electrical activity.
4. The human body can be assumed to be spherical in shape.
5. The electrical conductivity of the body is uniform throughout.

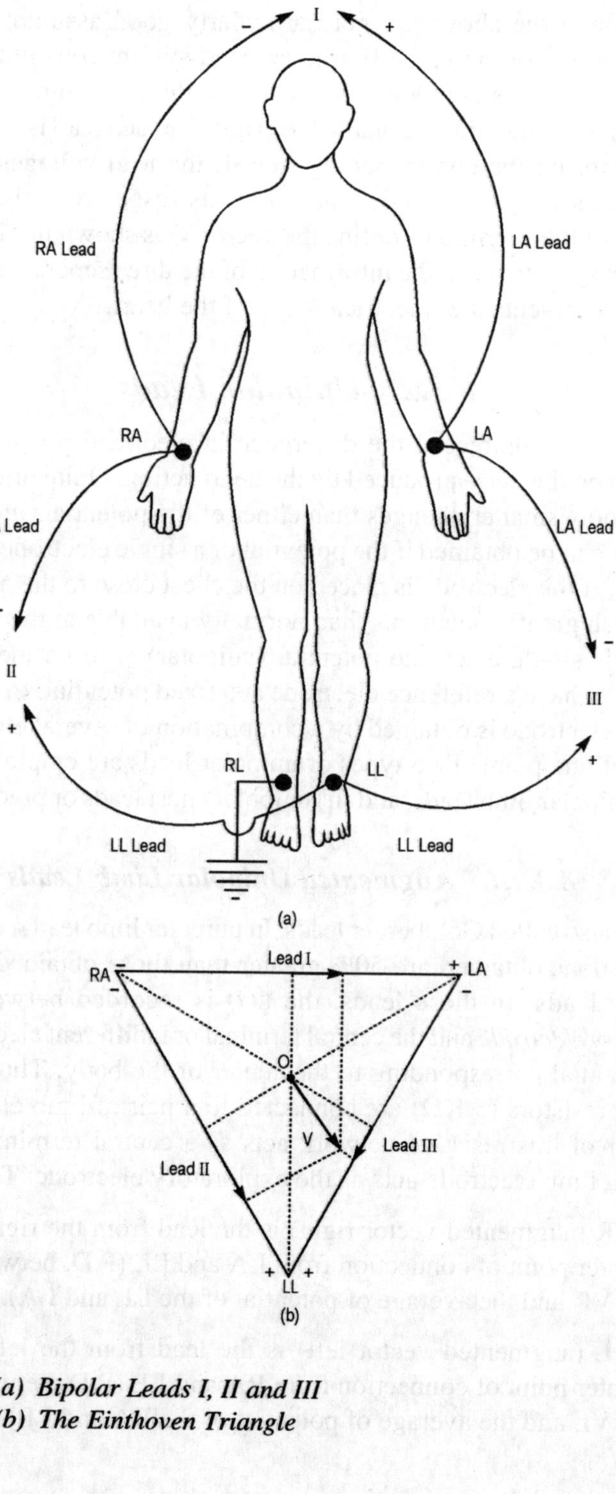

Figure 8.4 *(a) Bipolar Leads I, II and III*
(b) The Einthoven Triangle

Although the above are not particularly good assumptions from a theoretical point of view, the Einthoven lead system constitutes the most common clinical system today. According to this, the origin of the vector is at the centre of the triangle, while the origin for each lead is the projection of this vector on the corresponding side. If the lead voltages are plotted along the sides of the triangle, each from its respective origin, then the projection of their terminals define the vector V as shown in Figure 8.4(b). The vector V contains all the information of the three separate lead components and represents the 'electrical axis' of the heart.

8.3.2 Unipolar Leads

The standard leads record the difference in electrical potential between two points on the body produced by the heart action. Quite often, this voltage will show smaller changes than either of the potentials and, so, better sensitivity can be obtained if the potential of a single electrode is recorded. Moreover, if the electrode is placed on the chest close to the heart, we can detect much greater potentials than normally available at the limbs.

If this single electrode potential (unipolar) is to be measured, it is necessary to have a reference electrode at a fixed potential. In practice, the reference electrode is obtained by a combination of several electrodes tied together at one point. Two types of unipolar leads are employed: (i) augmented unipolar limb leads, and (ii) unipolar chest leads or precordial leads.

8.3.2.1 Augmented Unipolar Limb Leads

These are also called Goldberger leads. In unipolar limb lead measurements, the deflections obtained are 50% greater than those obtained in standard extremity leads. In these leads, the ECG is recorded between a single *exploratory electrode* and the central terminal or indifferent electrode, which has a potential corresponding to the center of the body. Thus, two equal and large resistors (5 KΩ) are connected to a pair of limb electrodes and the center of this resistive network acts as a central terminal, while the remaining limb electrode acts as the exploratory electrode. Thus,

- aVR (augmented vector right) is the lead from the right arm to the center point of connection from LA and LL (P.D. between RA, that is, VR and the average of potential of the LL and LA).
- aVL (augmented vector left) is the lead from the left arm to the center point of connection from RA and LL (P.D. between LA, that is, VL and the average of potential of the RA and LL).

- aVF (augmented vector foot) is the lead from the left leg to the center point of LA and RA (P.D. between LL, that is, VF and average of potential of the LA and RA).

This is illustrated in Figure 8.5(a). It should be noted that, in these three measurements, an indifferent electrode is formed at the negative input of the amplifier in all cases. The ECG waveform will thus be positive for aVL and aVF, but negative for aVR.

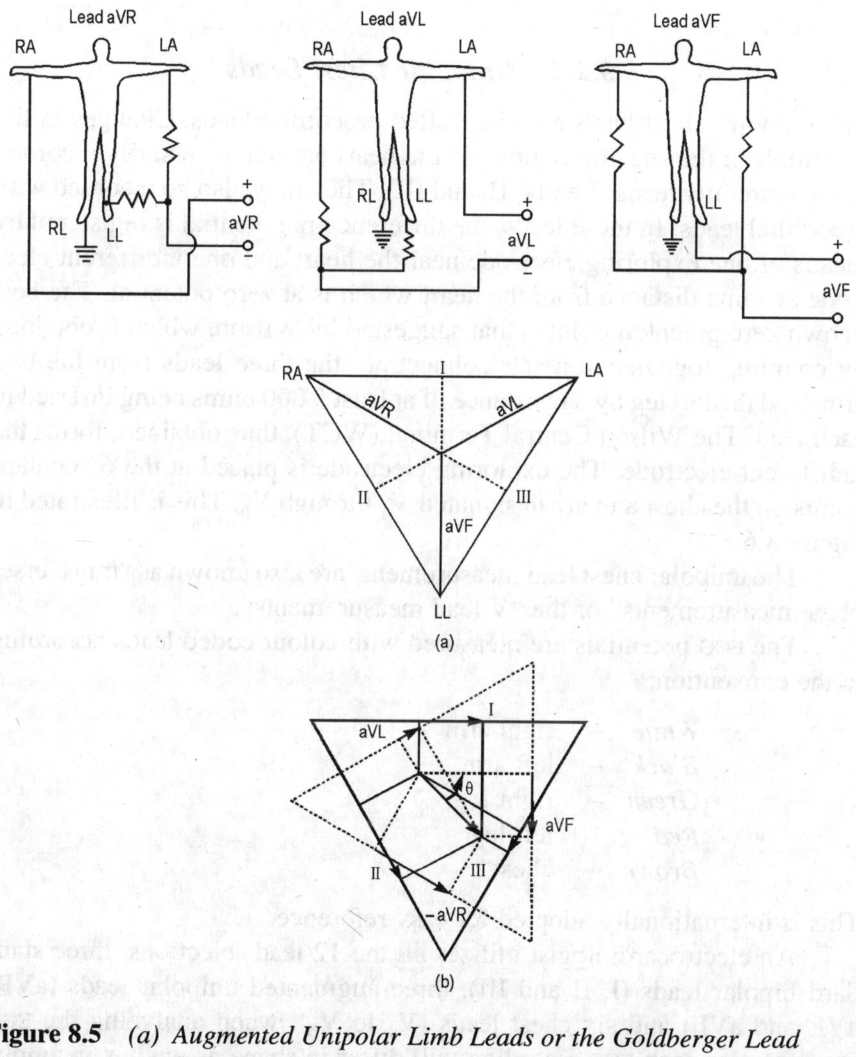

Figure 8.5 *(a) Augmented Unipolar Limb Leads or the Goldberger Lead System*
(b) Relationship of Frontal Plane ECGs to the Frontal Plane Axes of the QRS Complex

The unipolar limb lead electrode configurations are simply a projection of the same frontal-plane cardiac vector onto three different 60° axes rotated through 30° from the Einthoven configuration, as shown in Figure 8.5(b). The three unipolar leads bear a direct vector relationship to the three bipolar standard limb leads.

$$\text{aVR} = -\frac{\text{I}+\text{II}}{2};\quad \text{aVL} = \frac{\text{I}-\text{III}}{2};\quad \text{aVF} = \frac{\text{II}+\text{III}}{2}$$

8.3.2.2 Unipolar Chest Leads

The unipolar chest leads are also called precordial leads. Changes in the potentials in the specific regions of the heart are not necessarily recorded in the extremity leads (Lead I, II, and III). They may also be recorded with precordial leads. In these leads, the difference in potential is registered by means of one exploring electrode near the heart and one indifferent electrode at some distance from the heart which is at zero potential. The best known zero potential point is that suggested by Wilson, which is obtained by coupling together in a star connection, the three leads from the two arms and the left leg by a resistance of at least 5 000 ohms being inserted in each lead. The Wilson Central Terminal (WCT), thus obtained, forms the indifferent electrode. The exploring electrode is placed at the 6 standard points on the chest and are designated V_1 through V_6. This is illustrated in Figure 8.6.

The unipolar chest lead measurements are also known as "transverse-plane measurements" or the "V lead measurements".

The EGG potentials are measured with colour coded leads according to the convention:

White – right arm
Black – left arm
Green – right leg
Red – left leg
Brown – chest

This is internationally adopted for easy reference.

An electrocardiologist utilizes all the 12 lead selections: three standard bipolar leads (I, II and III), three augmented unipolar leads (aVR, aVL and aVF) and six chest leads (V_1 to V_6), when analyzing the EGG waveform as each ECG recording will differ in shape as well as in amplitude. Certain complexes within the EGG waveform are more clearly visible in one configuration than in another. The EGCs recorded from these 12 lead

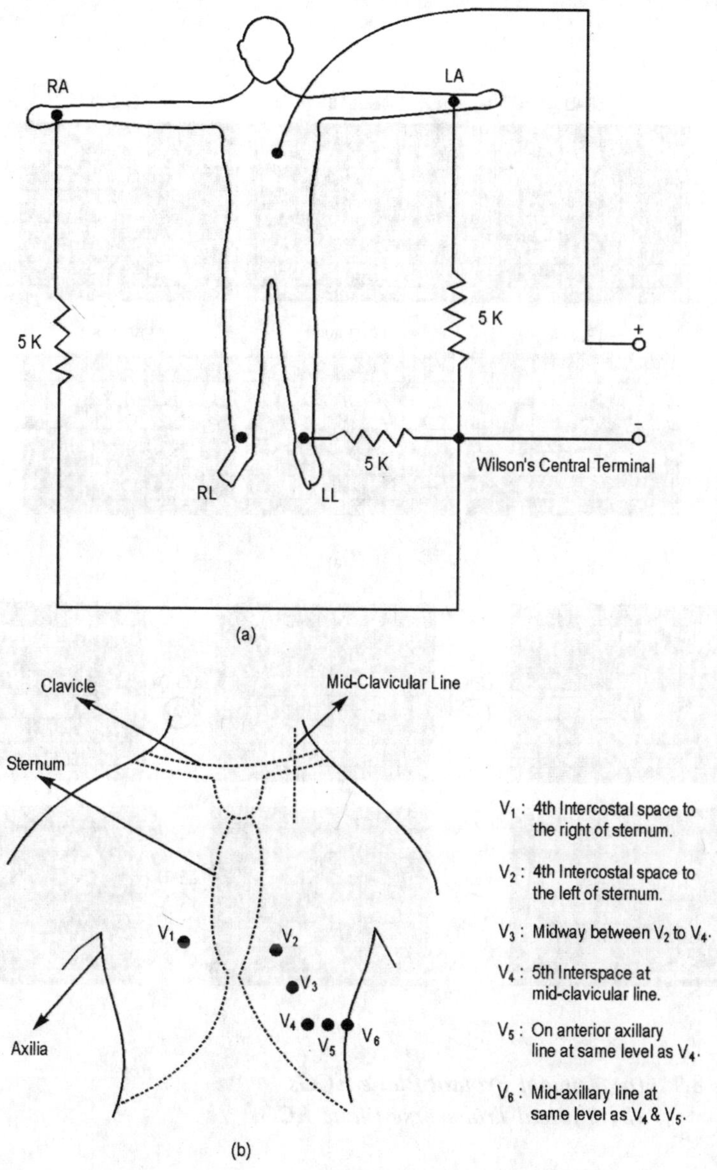

Figure 8.6 *(a) Connections of Limb Electrodes to Obtain Wilson's Central Terminal*
(b) Positions of Precordial Leads on Chest

selections are shown in Figure 8.7. It can be seen that the trace lead selection I or II most closely resembles the idealized waveform of Figure 8.6; some of the other traces are quite different in appearance.

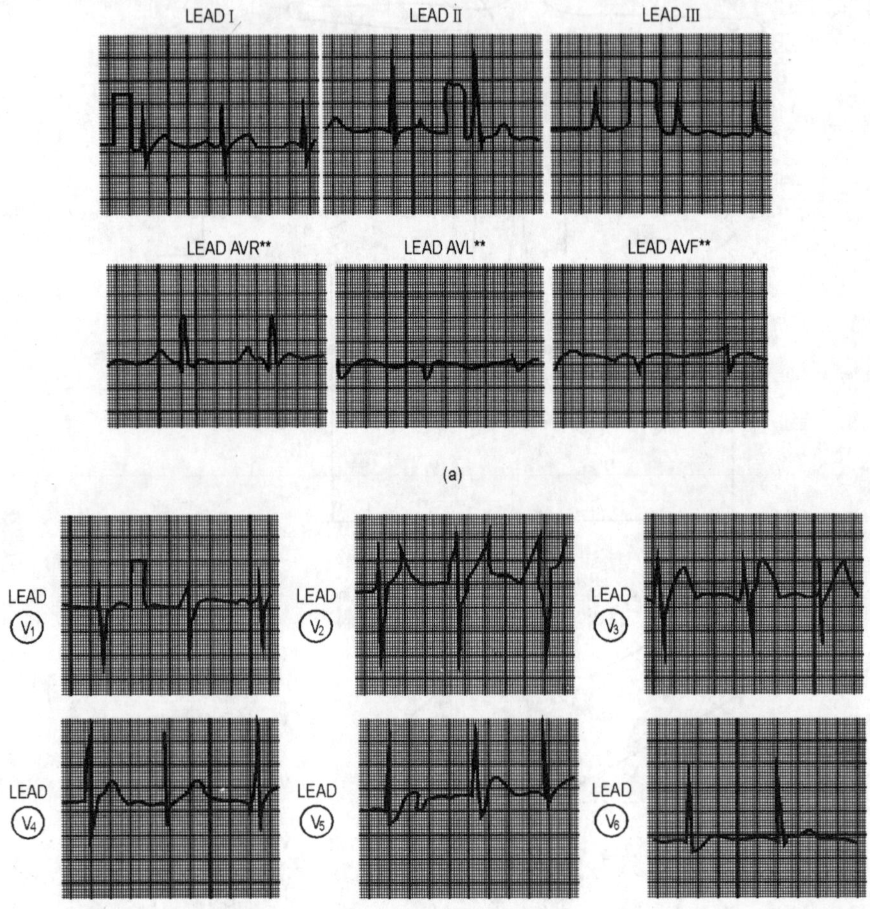

Figure 8.7 *(a) Typical Frontal Plane ECGs*
(b) Typical Transverse Plane ECGs

8.4 Determination of the Heart Rate

The ECG is usually and conventionally recorded at a paper speed of 25 mm/s. At this paper speed, five large squares represent 1 second, and fifteen large squares represent 3 seconds. Most recording papers have every fifteenth large square—a 3 second interval marked with a vertical line on the upper border.

The cardiac rate per minute may be estimated as follows:

1. The number of R-R or P-P intervals are counted in a 3-second period and the figure multiplied by 20.
2. The number of R-R or P-P intervals are counted in a 6-second period and the figure multiplied by 10.
3. The cardiac rate per minute may also be estimated by dividing the figure of 6 000 by the R-R or P-P interval (measured in a hundredth of a second). For example, if the R-R interval measures 0.08 second, the rate is 6000 ÷ 80 = 75 per minute.

8.5 Sagittal Plane ECG Measurements

The measurements of the cardiac vector projection in the sagittal plane are referred to as "the unipolar esophageal lead ECG measurements" or the "E lead ECG measurements." Although the unipolar esophageal lead ECG is rarely recorded these days, it can be recorded by attaching an esophageal lead to the positive input of the amplifier and forming an indifferent negative electrode using both arms and the left leg as with the transverse-plane V leads. Esophageal lead measurements are recorded from within the esophagus using a catheter or fine rubber tube, threaded with a wire with an electrode attached to its tip, as shown in Figure 8.8.

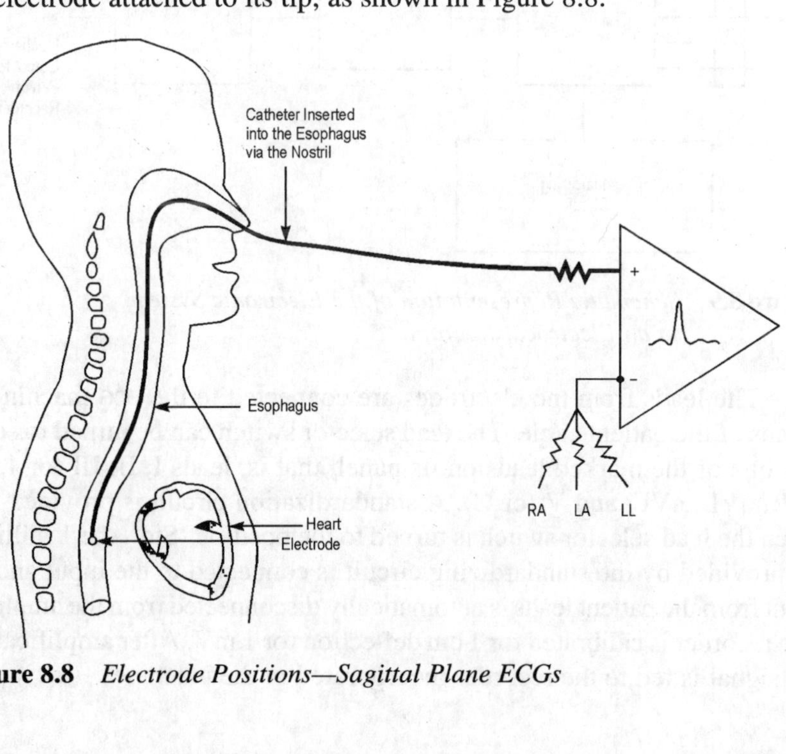

Figure 8.8 *Electrode Positions—Sagittal Plane ECGs*

8.6 Electrocardiograph

The electrocardiograph is an instrument which records the electrical activity of the heart. Electrical signals from the heart characteristically precede the normal mechanical function and the monitoring of these signals has great clinical significance. The ECG provides valuable information about a wide range of cardiac disorders, such as the presence of an inactive part (infarction) or an enlargement (cardiac hypertrophy) of the heart muscle. Electrocardiographs are used in the catheterization laboratory, coronary care units and for routine cardiac diagnostic applications in cardiology.

A block diagram of the electrocardiograph is shown in Figure 8.9. This "ECG machine" is supplied with 4 or more electrodes (normally made of silver) in the form of flat plates which can be strapped on to the hands and legs with rubber bands. Lead wires from the ECG machine can be plugged and screwed on to these electrodes. A special chest electrode is provided with rubber suction cup so that this chest electrode (or precordial electrode) can be clamped on the appropriate site (V_1 to V_6) and held by the partial vacuum created in the suction cup.

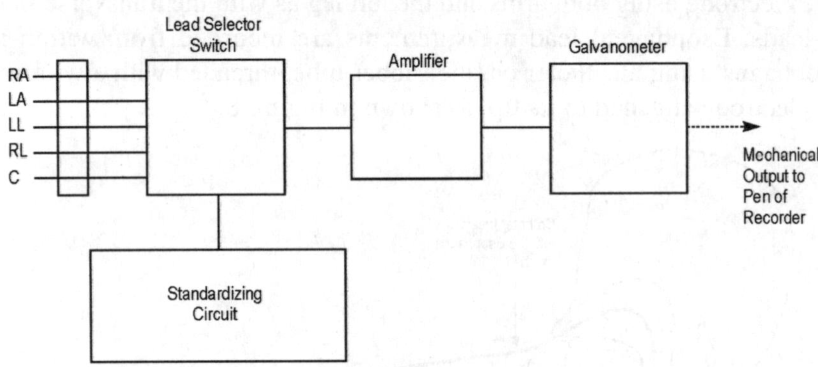

Figure 8.9 *Schematic Representation of the Electronic System in an Electrocardiograph*

The leads from the electrodes are connected to the ECG machine by means of the patient cable. The lead selector switch can be turned to select any one of the marked leads on its panel, that is, leads I, II, III (or 1,2,3) aVR, aVL, aVF, and V (or C). A standardization circuit is provided and, when the lead selector switch is turned to the position 'Std', the 1 millivolt DC provided by the standardizing circuit is connected to the input and the input from the patient leads is automatically disconnected from the amplifier. The recorder is calibrated for 1 cm deflection for 1 mV. After amplification, the signal is fed to the recorder incorporated in the machine.

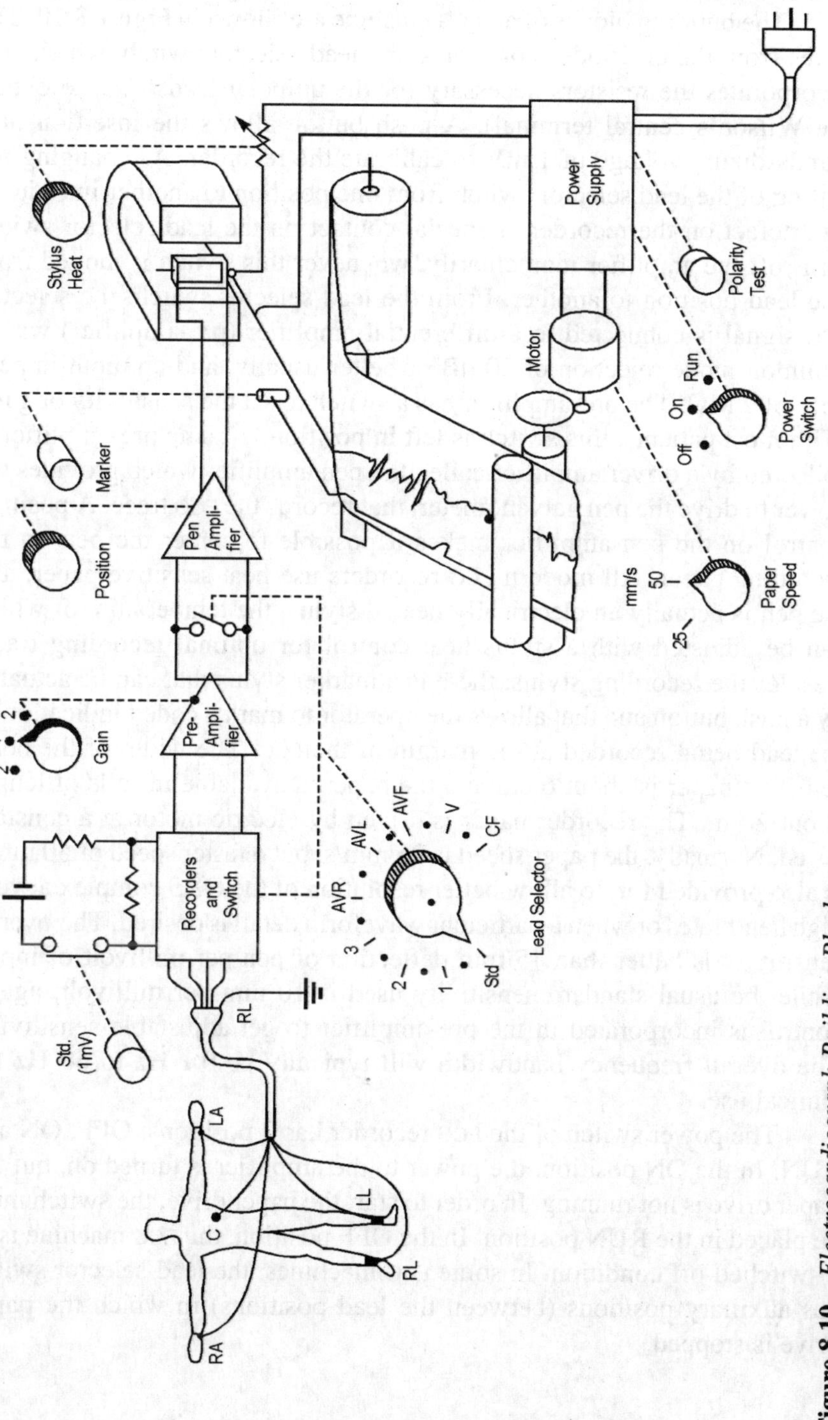

Figure 8.10 *Electrocardiograph Building Blocks*

The building blocks of an ECG machine are shown in Figure 8.10. The wires from the electrodes connect to the lead selector switch, which also incorporates the resistors necessary for the unipolar leads (*i.e.*, to obtain the Wilson's central terminal). A push button allows the insertion of a standardizing voltage of 1 mV to calibrate the recorder. As changing the setting of the lead selector switch from one position to another introduces an artefact on the recorder, a special contact on the lead selector switch turns off the amplifier momentarily, whenever this switch is moved from one lead position to another. From the lead selector switch, the selected ECG signal is connected to a differential amplifier (pre-amplifier) with a common mode rejection of 30 dB or better usually, and an input impedance of 2 MΩ. The pre-amplifier has a switch to set the sensitivity or gain. With most patients, this switch is left in position '1'. The pre-amplifier is followed by a driver amplifier called the pen amplifier which provides the power to drive the pen galvanometer, that records the ECG trace. A position control on the pen amplifier makes it possible to center the pen on the recording paper. All modern ECG recorders use heat-sensitive paper, and the pen is actually an electrically heated stylus, the temperature of which can be adjusted with a stylus heat control for optimal recording trace. Besides the recording stylus, there is a marker stylus that can be actuated by a push button and that allows the operator to mark a coded indication of the lead being recorded at the margin of the ECG. The width of the heat-sensitive paper is about 6 cm and the paper is available in rolls of length about 20 m. The recorder paper is driven by electric motor at a constant speed. Normally, the paper speed is 25 mm/s, but a faster speed of 50 mm/s is also provided for, to allow better resolution of the QRS complex at very high heart rates or when a particular waveform detail is desired. The overall sensitivity is better than 15 mm deflection of pen per millivolt of input, while the usual standard sensitivity used is 10 mm per millivolt; again, control is incorporated in the pre-amplifier to get adjustable sensitivity. The overall frequency bandwidth will typically be 0.1 Hz to 60 Hz for clinical use.

The power switch of the ECG recorder has 3 positions: OFF, ON and RUN. In the ON position, the power to the amplifier is turned on, but the paper drive is not running. In order to start the paper drive, the switch must be placed in the RUN position. In the OFF position, the ECG machine is in a switched-off condition. In some ECG machines, the lead-selector switch has auxiliary positions (between the lead positions) in which the paper drive is stopped.

Chapter 9

Electroencephalography

Electroencephalography is the study of the electrical activity of the brain. Usually this activity is recorded from electrodes placed on the scalp, although some relatively rare diagnostic procedures require electrodes on or beneath the cerebral cortex, and it has made many contributions to man's knowledge of the brain functions. For reasons which are examined later, it has been of greater help to neurology (the study of brain function) than to psychiatry (the study of mental processes).

9.1 Signal Sources

We have noted that many EEG signals are of microvolt levels and it should be remembered that the signal is arising not at the scalp but in the cerebral cortex, which is separated from the scalp by the cerebral spinal fluid (in which the brain is suspended) and by the skull. Parenthetically, we should note that engineers often suppose the skull to be an insulator because they usually see it dried and mounted. The living brain, however, is encased in living bone which is well permeated with conducting fluid. The amplifying system thus sees signals which arise in generators which have large, complex and variable source impedances. There may be large electrode offset potentials of the order of many millivolts developed between the electrodes and the scalp, unless a suitable electrode material is used. The high common-mode rejection ratio of the modern EEG amplifier will cancel the common-mode part of this signal but, in practice, small movements of the subject's head can cause substantial variations in the standing potential and if these are different in each lead, they will of course appear as differential signals.

Originally, it was thought that brainwaves represent a summation of the action potentials of the neurons in the brain. Now, it is believed that the electrical patterns obtained on the surface of the skull are the result of the graded potentials on the dendrites of neurons in the cerebral cortex and other parts of the brain, as they are influenced by the firing of other neurons that impinge on these dendrites. *Graded potentials* are variations around the average value of the resting potential. Thus, the EEG potentials originate within the dendrite potentials. Electric charges are transferred between one nerve fibre and another through a dendrite of a post-synaptic neuron during the release of *acetylcholine*. A great number of these potentials are then summed up to produce the EEG rhythms.

9.2 Input Electrodes

The input electrodes are the most critical components of the recording chain. To be of use in a routine EEG recording they must be small, easily affixed to the scalp with minimal disturbance of coiffure, cause no discomfort and remain in place for extended periods of time. They must also have some fairly rigid electrical specifications if the signals are to be recorded with acceptably low levels of distortion.

The most widely used electrodes are small silver pads electrolytically coated with silver chloride and attached to the scalp with a quick drying adhesive, usually *collodion*. A harness of rubber straps is also used to hold the electrodes in place. Before the electrodes are applied, the scalp area is degreased and cleaned with alcohol and the surface resistance reduced by the use of a conducting paste. Generally, this resistance range is from a few thousand ohms to nearly 100 KΩ, depending on the type of electrode used. These electrodes are satisfactory for most recordings in the range of 1 to 60 Hz.

9.3 Placement of Electrodes

The placement of electrodes on the scalp is commonly dictated by the requirements of the measurement to be made. In clinical practice, a standard pattern called the *10–20 electrode placement system,* is generally used. This system, devised by International Federation of Societies for Electroencephalography (IFSE), is so named because electrode spacing is based on intervals of 10 and 20% of the distance between specified points on the scalp. This gives adequate coverage of all the parts. This method of measurement is as shown in Figure 9.1.

a. Anterior–Posterior (Front–Back) measurements: The distance between the Nasion and Inion over the head in the midline is divided into the following five points:-

1. Frontal pole — FP - 10% of Nasion–Inion distance above Nasion.
2. Frontal — F - 20% of Nasion–Inion distance from FP.
3. Central — C - so on.
4. Parietal — P - so on.
5. Occipital — O - 10% of Nasion–Inion distance from Inion.

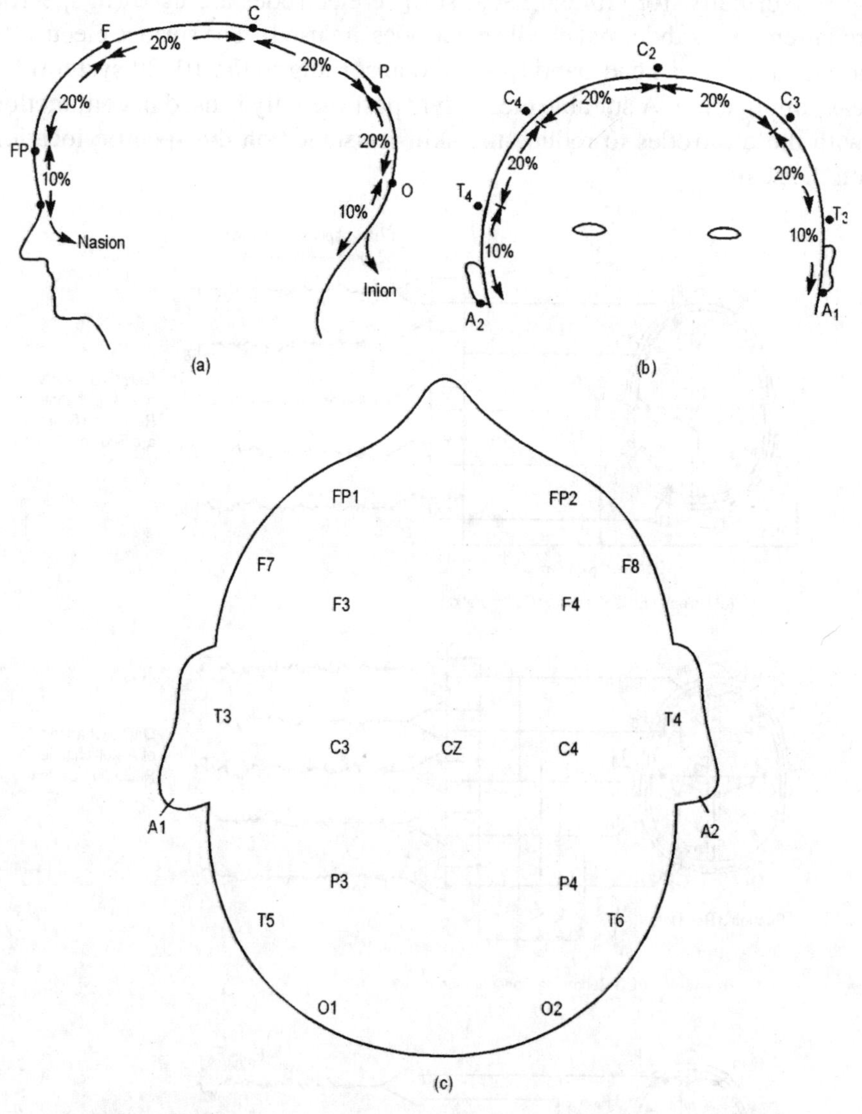

Figure 9.1 *The 10–20 EEG Electrode Placement System*

b. Lateral measurements: The distance is measured from the left to right pre-auricular (ear) points, passing through predetermined central points on the head.

1. Temporal points – T - 10% of this distance from the pre-auricular (ear) point on either side.
2. Central points – C - 20% of this distance.

Normally, for clinical EEGS, surface electrodes are used (in specific requirements, subdermal needle electrodes are used). The surface electrodes are mounted in a head-band (prepared according to the 10–20 system) for easy application. A suitable electrolyte paste or jelly is used in conjunction with the electrodes to reduce the skin resistance on the specific location on the scalp.

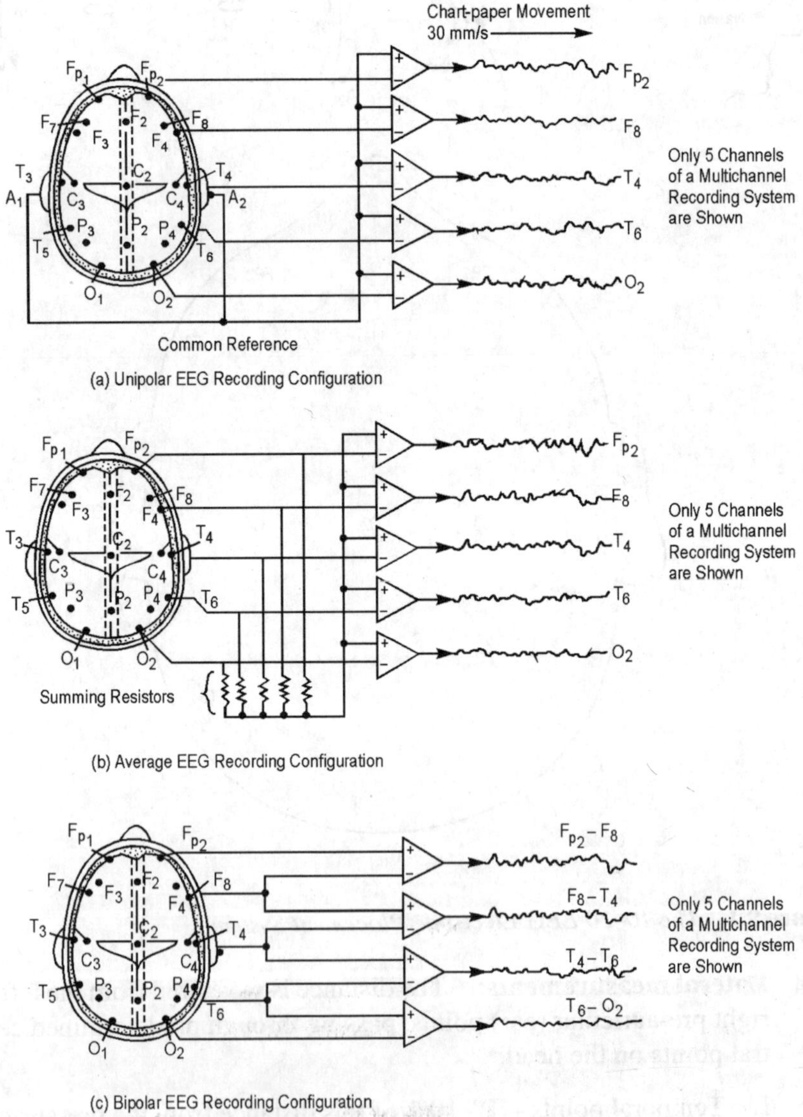

(a) Unipolar EEG Recording Configuration

(b) Average EEG Recording Configuration

(c) Bipolar EEG Recording Configuration

Figure 9.2 *EEG Recording Modes*

9.4 EEG Recording Modes

Three modes of recording or electrode connections used are shown in Figure 9.2:

1. Monopolar or unipolar (between one electrode and the distance reference electrode) and
2. Average mode or Wilson technique (between one electrode and the average of all other electrodes).
3. Bipolar (between each of a pair).

In the monopolar technique, Figure 9.2(a), the potential of each electrode is measured with respect to a reference electrode (common to all channels) attached to the ear-lobes or nostrils. Although one electrode is common to all channels, to reduce interference and artefacts, it is desirable not to ground this common electrode and so a separate ground electrode is often connected between the subject and the instrumentation ground.

In the average mode recording technique, Figure 9.2(b), the potential is measured between one of the electrodes (exploring electrode) and the central terminal which is formed by connecting all electrodes through high, equal resistors to a common point of a summing network, forming the system reference.

In the biopolar technique, Figure 9.2(c), the channels are connected in series between electrode pairs, and the potential difference is measured between successive pairs of electrodes (closely spaced). Thus, this does not measure far-field activity common to both electrodes, thereby obtaining a sharp localization of the response. Multichannel electrode encephalographs have (as many as the channels permit simultaneous recording from) several pairs of electrodes, reducing the total time required to complete the recordings. Eight channel recorders are very popular.

9.5 Electroencephalograph

The electroencephalograph is provided with two types of electrodes. One set is for placement on the ear-lobes. The electrode itself is typically an 8 mm diameter silver disc. Teflon clips about 5 mm long are fixed to the electrode so that it can be clamped to the ear. The other types of electrodes provided are for placement on the scalp. They are small disc electrodes fixed to the scalp with paste. A thin layer of electrode jelly is placed between the scalp and the electrode to improve conductivity.

In electroencephalography, a large number of electrodes are placed on the head and most of them are used for recording at any one time. Therefore, electroencephalographs must have a large number of channels.

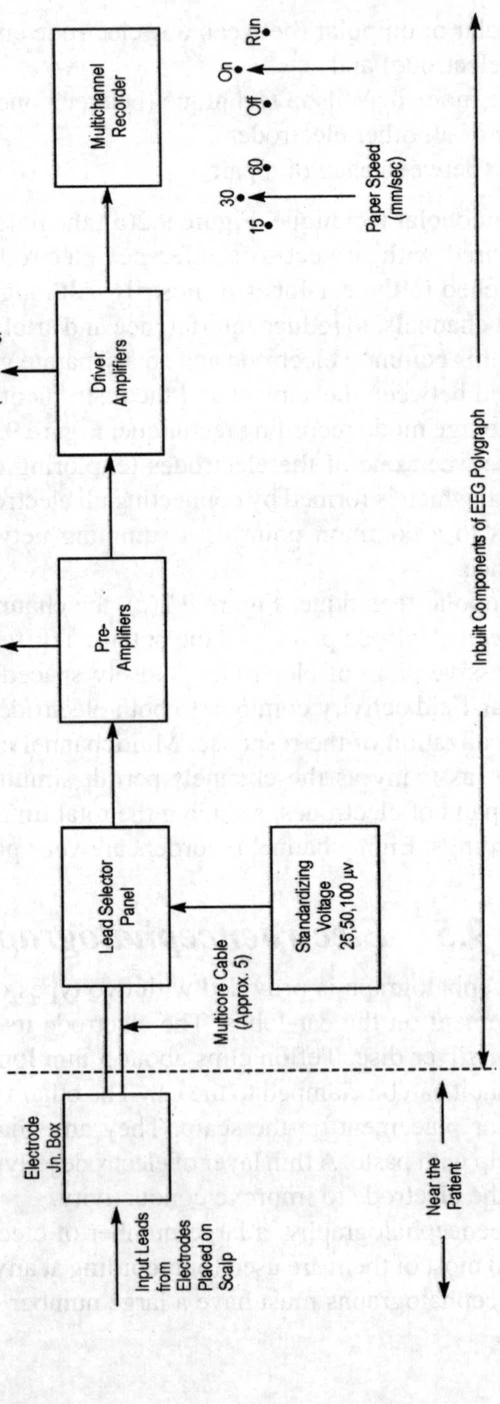

Figure 9.3 *Block Schematic Diagram of an EEG Recording System*

The common figures are 9, 12, 16 and, at times, as many as 32 channels are provided. In a typical 9 channel electroencephalograph (Figure 9.3), 8 channels are used for EEG patterns. The 9th channel is a general purpose channel and can be used for ECG and EMG as well as EEG. With such an electroencephalograph usually 21 electrodes are placed on the scalp and 2 more electrodes, one on each ear-lobe. The leads from each of the electrodes, constituting the patient cable, are brought over to the junction box.

As in the electrocardiograph, there is a selector switch. However, since there are 8 channels, the selector switch is more complex and selects a pattern instead of just two active leads. For each channel two active leads are selected so as to get electrical activity from different locations of the brain. The selector switch has more than one position so that different patterns can be selected. Either a bank of rotary switches or a panel of push buttons is used. The selector switch also permits one of several calibration signals to be connected to all the channels for calibration of the entire instrument. The calibration signal is usually a DC voltage of 25, 50 and 100 μV to accommodate a wide range of EEG signals.

The EEG signals are then amplified by means of high quality differential amplifiers with good CMRR (about 80 dB) and an input impedance greater than 10 MΩ, having a (cascaded) gain of the order of 10^6. The differential pre-amplifier is generally followed by a power amplifier to drive the pen mechanisms for each channel. In nearly all clinical instruments, the amplifiers are AC coupled with low frequency cut-off below 1 Hz and a bandwidth extending to somewhere between 50 and 100 Hz. Most modern electroencephalographs include upper and lower frequency limits to allow the operator to select a bandwidth suitable for conditions of the measurement. In addition, some instruments include a fixed 50 Hz notch (rejection) filter to reduce the power line interference. Filters also are provided to cut off higher frequencies, particularly those due to electromyogram signals. The filters are in-steps so that the cut-off frequency can be selected.

The read-out in a clinical electroencephalograph is a multichannel pen recorder with a pen for each channel. The standard speed is 30 mm/s. EEG machines are also provided with chart speeds of 15 mm/s to conserve recording paper during set up time and 60 mm/s for improved detail of higher frequency signals. The paper used for recording EEG is generally 25 cm wide (in folded form). A resistance measuring ohmmeter is incorporated in the EEG machine so that the electrode resistance can be checked conveniently. Some electroencephalographs have a provision for connecting an automatic brainwave analyzer to the machine. The provision is simply an output jack connected to the output of the pre-amplifier or main amplifier.

9.6 Application of the EEG

The EEG is primarily used in clinical neurology for partially assessing a subject's neurological state; some applications are as listed below.

(a) **Anesthetic level:** As well as its utility in the neurological clinic in surgery, a modified EEG recording using three leads is sometimes made in order to help the anesthesiologist determine the depth or intensity of anesthesia. Some workers have used the EEG signal in a closed loop controller to keep a constant anesthetic level.

(b) **Epilepsy:** The most common condition in which EEG is valuable as a diagnostic tool is in epilepsy. The EEG is of great help in determining the acuteness of epilepsy and is versatile in establishing the optimum course of treatment.

(c) **Brain injury:** For example, an injury to a specific region of the brain can leave a permanent scar on the cerebral cortex. Such scar tissue is electrically inert but has an irritative effect on the nearby healthy cortex. The EEG will often show a localized spike discharge and will suggest, among other possibilities, surgical removal of the damaged tissue.

(d) **Tumor location:** When the brain is invaded by some forms of tumor, a considerable portion of the active nervous tissue may be displaced by electrically inert new growth. The expanding new growth can interfere with the blood supply to the neighbouring areas and a consequent malfunctioning of nerve cells around, it manifests by a large, slow discharge—the delta rhythm. Thus, the EEG can be used in locating the tumors within the brain.

(e) **Alertness monitor:** The EEG is also a very subtle estimator of the differences between sleep and wakefulness. A number of sleep states can be distinguished. For those who must be alert at specified times during a long task, piloting a spacecraft is a case in point, the EEG can be made the basis of a reliable "state of alertness" monitor.

(f) **Monitoring during surgery:** In many surgical procedures involving the heart, the ECG waveform cannot be monitored, thus the EEG signal is used as an indication of the subject's well-being.

(g) **Efficacy of yoga:** EEG recordings made on a person during yoga and comparing the same before going into yoga, could give us a concrete idea regarding the efficacy of yoga.

Chapter 10

Electromyography

While the myograph records the mechanical effects of a muscular contraction, the electromyograph records the electrical effects of such contractions. Muscular contractions are caused by the depolarization of muscle fibres. This depolarization produces an action potential as mentioned in Chapters 2 and 3, this muscular action potential being known as an electromyogram or EMG. An EMG will be produced in a muscle when the muscle contraction is caused either by voluntary muscle action or by electrical stimulation of the muscle.

10.1 EMG Electrodes

Two main types of electrodes are used. They are: the surface electrodes and needle electrodes.

10.1.1 Surface Electrodes

Surface electrodes are usually used where a broad or global pick-up of a number of muscles or a large area of muscle is desired. Surface electrodes will miss deeper potentials which the needle electrodes pick up very well. The needle electrodes do not miss any of the potentials that the surface electrodes immediately overlying them do miss. Surface electrodes consist of metal discs (usually made of silver) varying in diameter from 0.5 to 2.5 cm. In practice, the potential difference (P.D.) between two suitably placed electrodes is recorded by connecting each to one side of the differential amplifier, a third electrode being used to connect the patient with the ground.

10.1.2 Needle Electrodes

They are inserted into, or close to, the tissue from which electrical activity is to be recorded. Needle electrodes will record activity from a much smaller area than surface electrodes and, therefore, permit the activity of single motor unit to be studied. The outer shaft of the needle electrode is usually made of stainless steel, because of its mechanical strength. The recording area of needle electrodes is usually less than 1 sq mm. Some of those used in clinical practice are illustrated in Figure 10.1.

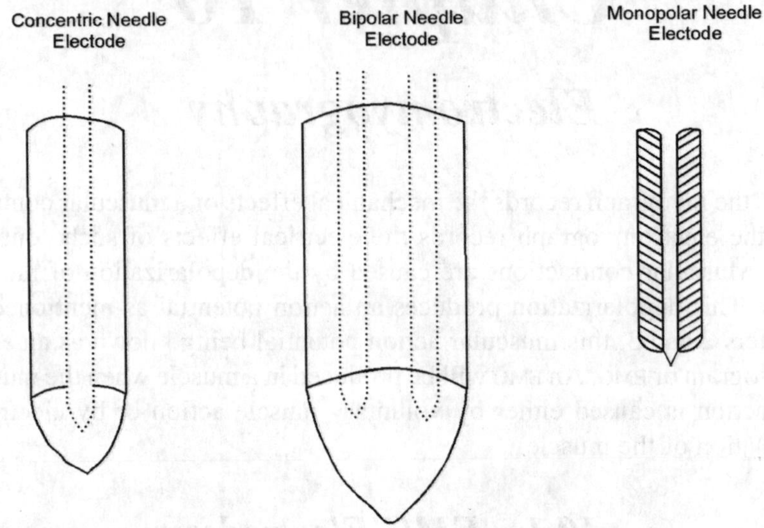

Figure 10.1 *Needle Electrodes used in EMG*

The concentric needle electrode consists of a hollow steel needle through which runs a silver, steel or platinum wire, insulated except at its tip. The P.D. between the outer shaft and the tip of the wire is measured. The patient is grounded by a separate surface electrode. The bipolar needle electrode is a hollow needle containing two pieces of platinum wire each of which is insulated, except at its tip. The outer shaft is grounded and the P.D. between the two inner wires is measured. The monopolar needle electrode is a solid, stainless steel needle which is insulated except at its tip. The P.D. between this needle and a distant surface electrode, or from a second needle electrode, is measured and the patient is grounded separately.

10.2 The Electromyograph System

The basic arrangement of a typical electromyograph system is shown diagramatically in Figure 10.2. EMG potentials from the tissue under study are picked up by the recording electrodes and fed to an amplifier which is connected to a CRO, a loudspeaker system through an audio amplifier, and a magnetic tape recorder. The amplified signals can thus be monitored visually and accoustically, and stored for further analysis at a later date.

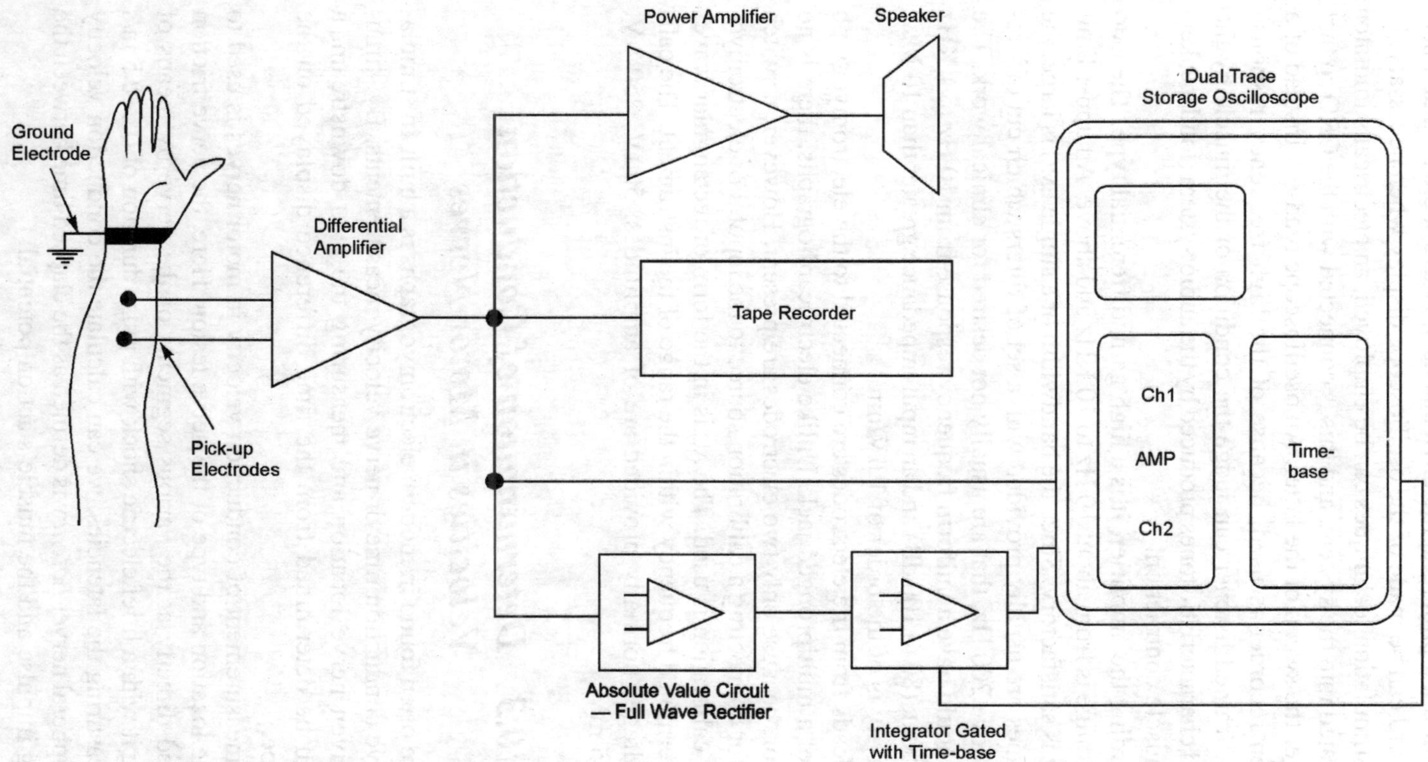

Figure 10.2 *An Electromyograph System*

In addition, a photographic device is usually available so that a permanent record can be made of the oscilloscope display when necessary. In order that stimulation techniques can be employed, an electrical stimulator is incorporated into the system, and this is connected with the CRO so that it can trigger the sweep of the latter. An oscilloscope read-out instead of a graphic pen recorder is used, because of the high frequency response required. A trained listener can judge the condition of the muscle by the volume and characteristic tones produced by the audio system (loudspeaker) during a muscle contraction.

Regarding the amplifier, it is a high gain, differential type. The EMG frequency range is from about 10 Hz to 10 KHz and above. An upper limit of 1000 Hz is satisfactory. Since the bandwidth of a single EMG is large, the EMG machines are usually provided with a set of filters which cut off frequencies above 200 Hz, that are usually not desired for clinical work. The amplifier should have a uniform frequency response from 10 Hz to 1 KHz with high CMRR (80 to 100 dB) and an input impedance greater than 10 MΩ. The sensitivity is of the order of 1 μV/mm.

The leads from the electrodes are connected to the electromyograph by means of a multipronged jack. Unlike electrocardiographs, there is no lead selector, because only two electrodes are present. However, a switch is present for connecting a calibration source to the input. For electromyographs, AC calibration is used. The AC is in the form of rectangular waves of a predetermined frequency within the range of the instrument. The calibration voltages normally provided are of amplitudes: 50μV, 250 μV, 1 mV and 5 mV.

10.3 Determination of Conduction Velocities in Motor Nerves

One feature often found in modern electromyographs is a built-in stimulator for nerve conduction time or nerve velocity measurements. By stimulating a given nerve location and measuring the EMG downstream, a latency can be determined from the time difference displayed on the oscilloscope.

The measurement of conduction velocity in motor nerves is used to indicate the location and type of the nerve lesion. Here, the nerve friction is examined directly at the various segments of the nerve by means of stimulating it with a brief electric shock with pulse duration of 0.2–0.5 m/s and by measuring the latencies, we can calculate the conduction velocity in that peripheral nerve. *Latency* is defined as the elapsed time between the stimulating impulse and the muscle's action potential.

Figure 10.3 illustrates the measurement procedure. The EMG electrode and the stimulating electrode are placed at two points on the skin, separated by a known distance, l_1. A brief electrical pulse is applied

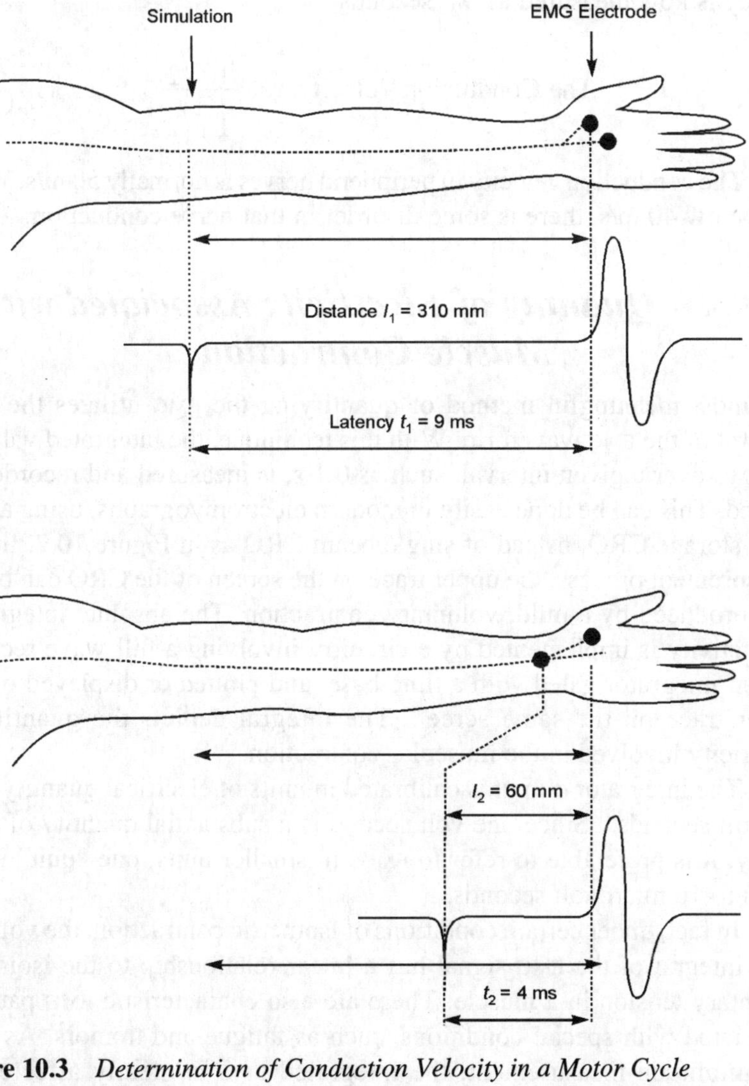

Figure 10.3 *Determination of Conduction Velocity in a Motor Cycle*

through the stimulating electrode. When the excitation reaches the muscle, it contracts with a short twitch. Since all the nerve fibres are stimulated at the same time and the conduction velocity is normally the same in all nerve fibres, there is a synchronous activation of the muscle fibre. This action potential of the muscle is picked up by the EMG electrode and is displayed on the oscilloscope along with the stimulating impulse. The elapsed time,

't_1' (latency), between the stimulating impulse and muscle's action potential is measured. Now, the two electrodes are repositioned with the distance of separation as l_2 metres. Among the distances l_1 and l_2, $l_2 < l_1$, the latency is now measured as 't_2' seconds.

$$\text{The Conduction Velocity, v} = \frac{l_1 - l_2}{t_1 - t_2} \quad \text{....(10.1)}$$

The conduction velocity in peripheral nerves is normally 50 m/s. When it is below 40 m/s, there is some disorder in that nerve conduction.

10.4 Quantity of Electricity Associated with Muscle Contraction

The most meaningful method of quantifying the EMG utilizes the time integral of the EMG waveform. With this technique, the integrated value of the EMG over a given interval, such as 0.1 s, is measured and recorded or plotted. This can be done easily in modern electromyographs, using a dual trace storage CRO instead of single beam CRO as in Figure 10.2. In this measurement process, the upper trace on the screen of the CRO can be the EMG produced by a mild, voluntary contraction. The absolute integral of this activity is implemented by a circuitry involving a full-wave rectifier and an integrator gated with a time-base, and plotted or displayed on the lower trace on the same screen. The integral depicts the quantity of electricity involved in the muscular contraction.

The integrator output is calibrated in units of electrical quantity, that is "volt seconds". Since one volt second is a substantial quantity of electricity, it is preferable to refer to EMGs in smaller units, one "unit" being equal to 10 microvolt seconds.

In fact, under certain conditions of isometric contraction, the voltage-time integral of the EMG signal has a linear relationship to the isometric voluntary tension in a muscle. There are also characteristic EMG patterns associated with special conditions, such as fatigue and tremors. As with the amplitude measurement, the integrated EMG is greatly affected by electrode placement, but with a given electrode location, these values provide a good indication of muscle activity.

Applications of EMG

Electromyographs are generally designed to have a wide range of uses in the field of electrophysiological testing with applications in physical medicine and rehabilitation, clinical neurophysiology and neurology, psychiatry, and in the teaching of electrophysiology.

10.5 Electroretinograph

The recording and interpreting the electrical activity of the eye is called electroretinography. All the sense organs are connected to the brain, but the eye has a special relationship, as the retina is an extension of the cerebral cortex. The potentials within the eye may be recorded relatively easily because of its exposed position. The cornea is about 20 mV positive, relative to the fundus of the eye. The fundus is the back of the interior of the eyeball. If the illumination of the retina is changed, the potential changes slightly in a complex manner. The recording of these changes is called the electroretinogram (ERG). The largest variations are usually slightly less than 1 mV.

Electrode Placement

A bipolar recording technique is used. The exploring electrode is placed on a saline-filled contact lens. A silver–silver-chloride electrode, embedded in the specially made contact lens, is in direct contact with the cornea and a 'common' silver–silver-chloride electrode is placed on the subject's forehead, as shown in Figure 10.4(a). The application of the contact lens electrode is painless and it is thus suited for clinical use.

Recording Techniques

The subject *is placed* in front of a light source which is usually a stroboscopic light, since this light can be adjusted to flash at various rates. When light falls on the retina, the absorption of photons by photopigments localized in the outer segment of the retina's photoreceptors is taking place. This causes the breakdown or bleaching of photopigments, which results in the liberation of ions that cause a change in the membrane potential. This, in turn, results in the development of action potential that is transmitted down the optic nerve. This action potential is picked up by the electrodes and fed to the bioamplifier and then to the recorder. The recording set-up is similar to the ECG recorder. Of importance is the fact that the eye

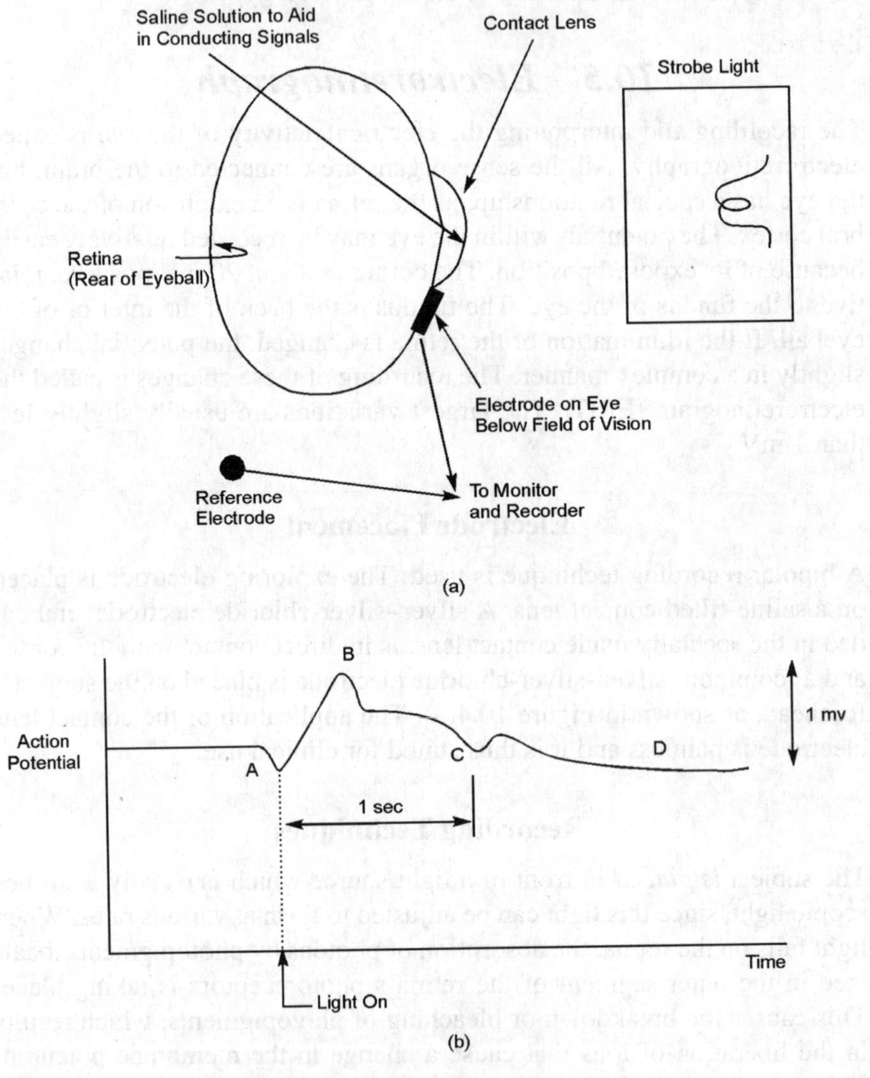

Figure 10.4 *(a) ERG—Electrode Placement System*
(b) Electroretinogram

is very sensitive to direct current (just as the heart has been found to be very sensitive to 50 Hz alternating current). Thus, in the equipment design, adequate capacitive coupling is provided in order to prevent any DC from reaching the eye. In other words, a capacitor may be put in series with each ERG electrode to eliminate DC currents.

Electroretinogram

Figure 10.4(b) shows a typical ERG. Before the flash of light incident on the eye, there is a constant horizontal line in the recorder. In response to a 2-s flash of light, an ERG is developed. Probably the curve originates from the pigment layer beyond the photoreceptors (extra-retinal). The first part 'A' of the response to a brief flash of light is due to the Early Receptor Potentials (ERP) generated by the incident light which induces changes in the photopigment molecules. The second component 'B' with a decay of 1 to 5 m/s is due to the Late Receptor Potentials (LRP) produced by the synaptic ending of the photoreceptors. This is the maximum output from the receptors. The 'C' wave, recorded at the offset of the light stimulus, is largely associated with the off-response of ERP and LRP.

ERGs are used for determining retinal degeneration. By calibrating the light source (amplitude and distance from the patient), the recorded signal can be diagnosed. The amplitude of the response signal is proportional to the intensity and duration of the light falling on the eye—the higher the response voltage, the better the condition of the retina.

10.6 Electroocculograph

Electroocculography is the recording of the biopotentials generated by the movement of the eyeball. A record of corneal–retinal potentials associated with eye movements is called the electroocculogram (EOG). This is simpler to record than the electroretinogram. The EOG potentials are picked up by small surface electrodes placed on the skin near the eye. One pair of electrodes is placed above and below the eye, to pick up the voltages corresponding to the vertical movements of the eyeball. Another pair of electrodes is positioned to the left and right of the eye, to measure horizontal movement. The recording pen is centred on the recording paper, corresponding to the voltages accompanying it. An alternating potential difference of 1 mV or more is recorded. The source of electrical energy in the eye is the corneoretinal potential or electrostatic field, which rotates with respect to the eye.

Some diseases which affect the steady potential of the eye can be usefully studied using the EOG. They are:

i. the effects of certain drugs on the eye movement system can be determined
ii. the state of the semicircular canals is analyzed by an EOG

iii. diagnosis of neurological disorders may be possible
iv. the level of anesthesia administered can be indicated by the characteristic eye movements.

10.7 Electrogastrograph

Electrodes appropriately inserted into the stomach, bladder, or placed on the surface of the body over these organs can serve to monitor the slowly varying potentials generated by the muscles in these organs. These potentials are known as "smooth muscle potentials". Although these potentials are rarely monitored, the recording of the electrical activity produced by the stomach has been termed the "electrogastrogram" (EGG).

The EGG is characterized by a slowly changing DC potential (below 1 Hz), which would normally be recorded on a DC coupled instrument at a sensitivity of 10 mV per division and at a sweep speed of perhaps 1 s per division.

Chapter 11

Respiration

The primary functions of the respiratory system are to oxygenate blood; that is, to dissolve oxygen into the blood and to remove carbon dioxide from the contents of the blood. If the blood is not oxygenated due to some failure of the circulatory system, the oxygen content of the blood will rapidly decrease and after about 60 to 90 seconds the subject will become unconscious, with death occurring in 4 to 5 minutes. The cessation of breathing is called *apnea*.

11.1 Mechanism

The lungs are the major component of the respiratory system and it is in the lungs that the oxygenation of blood occurs. When the lungs are forced to expand by the muscular contraction of the diaphragm and expansion of the thoracic cage by contraction of the rib muscles, air enters the lungs via the bronchi and is diverted to millions of small air sacs known as *alveoli*. The membrane comprising the alveoli is moist and the oxygen contained in the air is dissolved by this moisture. Interspersed with the alveoli are fine capillaries, branching from the circulatory system, through which blood is continually flowing. The oxygen dissolved in the moist surface of the alveoli diffuses into the blood stream via these capillaries. The carbon dioxide contained in the blood stream is also diffused through the alveolar membrane to be expelled along with the expired air. Some of the principal components of the respiratory system are shown in Figure 11.1.

A grown man at rest inhales about 500 cc of air with each breath and takes between 12 to 18 breaths per minute. During periods of moderate exercise, he may inhale one litre or more with each breath and take up to 25 breaths per minute. At maximum exercise, a vital capacity of 4 to 5 litres is reached.

The action of breathing is controlled by muscular action, causing the volume of the lung to increase and decrease. During normal breathing the lung does not contract to its minimum possible volume, nor does it expand to its maximum possible volume. Minimum lung volume is obtained when

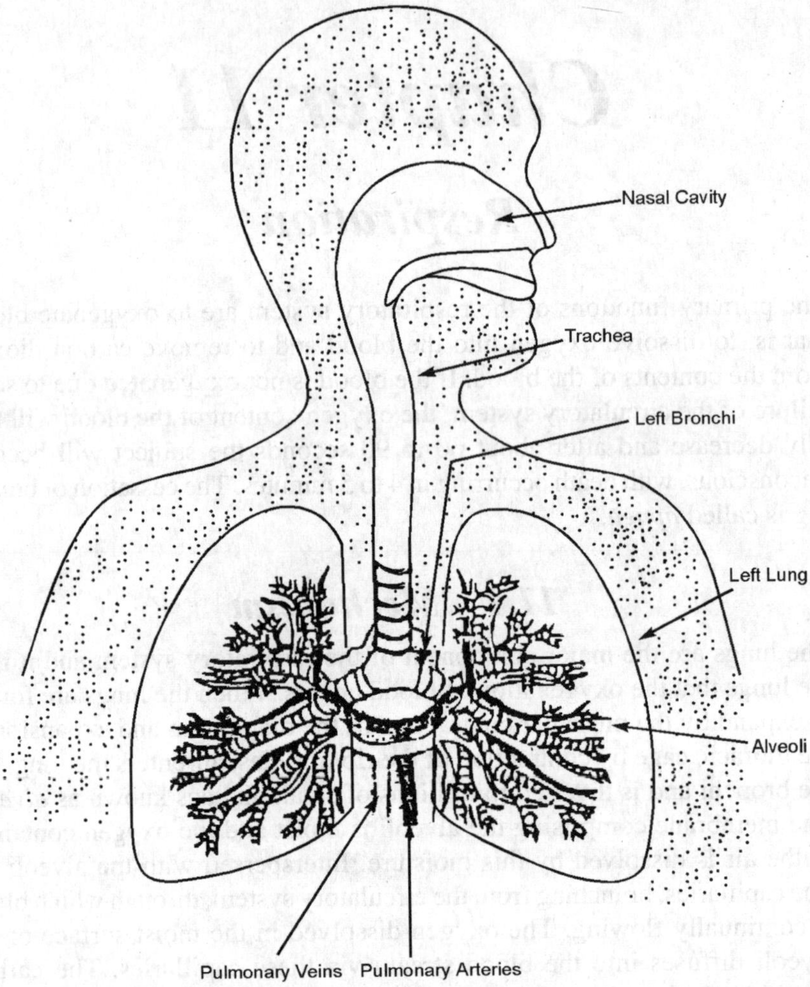

Figure 11.1 *Respiratory System*

exhaling with extreme effort to expel the maximum possible amount of air and maximum lung volume is obtained when inhaling with maximum effort. Figure 11.2 shows, diagramatically, the changing lung volume that may be expected for a resting man and the lung volume that may be achieved during maximum inspiratory and expiratory effort.

Air-flow volume in and out of the lungs is measured by a *spirometer*, and the recording of lung-volume changes with time is known as the *spirogram.* Instruments that simply detect respiratory activity are referred to as *pneumographs* and the resulting recording of respiratory-activity changes

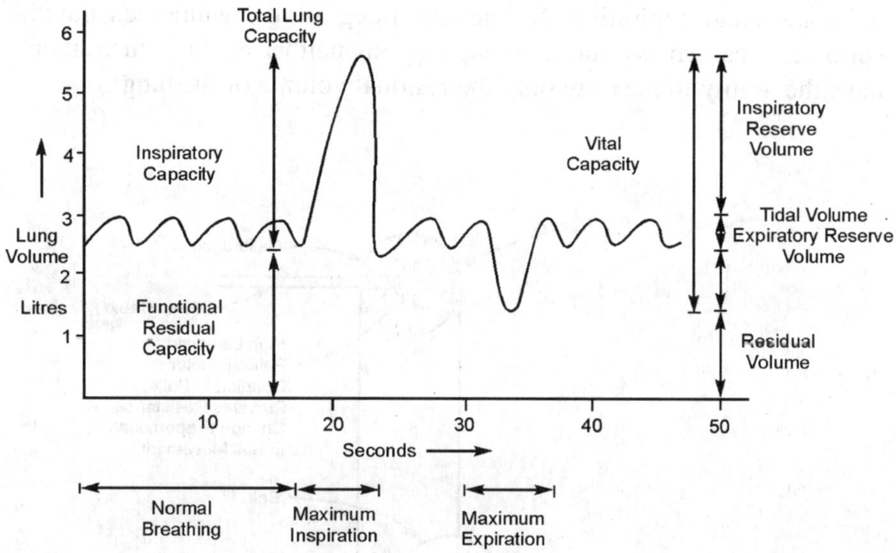

Figure 11.2 *Respiratory Changes in Lung Volume*

with time is known as a *pneumogram.* A spirogram is normally only required when attempting to analyze the respiratory system or to detect a malfunction of this system and is rarely used for routine monitoring. A pneumogram may be used for routine monitoring and is basically used to indicate the fact that the subject is breathing. Breathing rate may be obtained from either a spirogram or a pneumogram.

11.2 Spirometer

A conventional spirometer is shown in Figure 11.3. Referring to this device, inspiration and expiration raises and lowers a counterbalanced bell located in a container full of water. The movement of this bell is transferred to a pulley whose periphery contains a calibration of bell displacement which is, of course, related to the bell's air volume. Respiratory volume may be read directly from this calibrated pulley. This pulley may also be coupled to a high linearity potentiometer fed from a constant-current source; the output voltage will then be proportional to the changing resistance. The resistance change in this potentiometer is used to indicate respiratory volume. A spirogram recorded with this system is also shown in Figure 11.3. A spirometer is inherently a heavily damped device, containing appreciable hysteresis. So, small and subtle changes in inspiration and expiration volumes are not recorded with this device.

Figure 11.2 shows theoretical changes in total lung volume with inspiration and expiration. Neither the integrating pneumotach nor the spirometer can show total lung capacity, as neither of these instruments have the ability to measure only the residual volume of the lung.

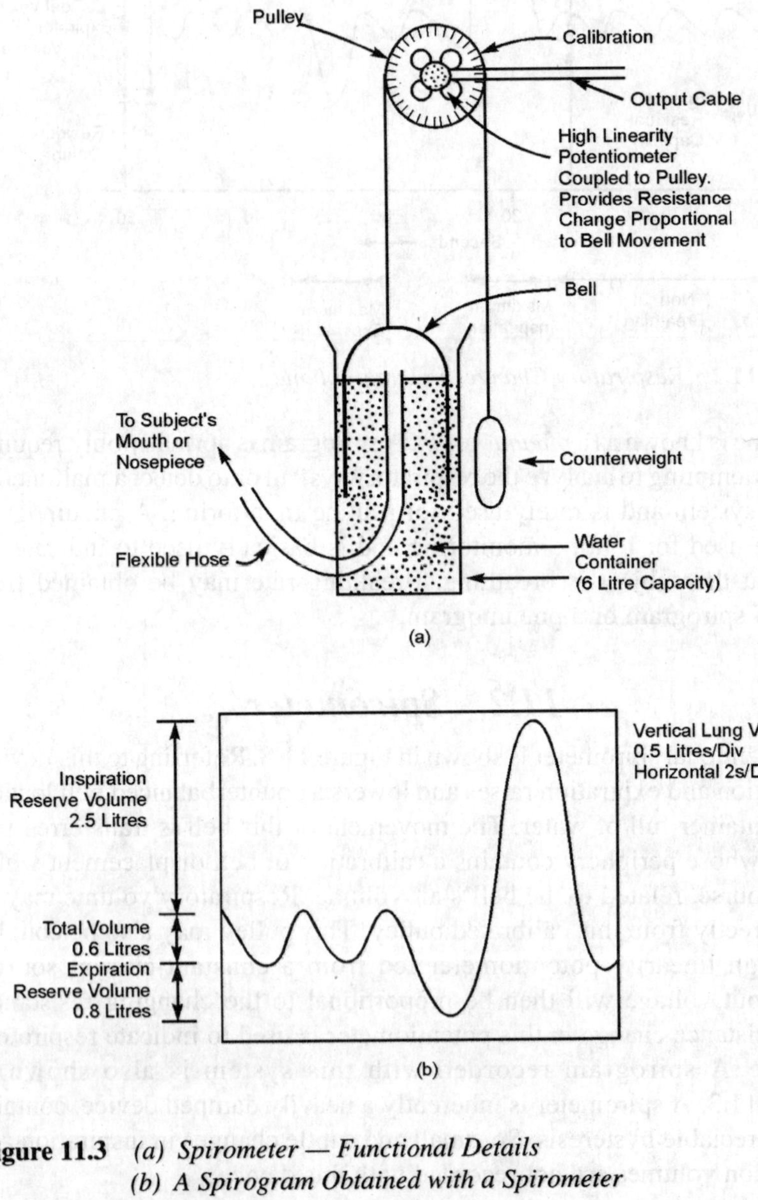

Figure 11.3 *(a) Spirometer — Functional Details*
(b) A Spirogram Obtained with a Spirometer

11.3 Pneumotachography

A method for determining the rate of air flowing into the lungs and by integrating this rate, determining the tidal volume rate, is called *pneumotachography*. This technique uses a differential pressure transducer, a device which has an electrical output only when changing pressure is sensed. By placing a mask over the patient's nose and mouth which allows normal breathing and placing a fine mesh screen in a tube leading from the mask to the surrounding air, a slight pressure differential will develop in the tube on either side of the screen (refer Figure 11.4). This pressure differential is found to be proportional to the air-flow velocity, and can be detected with the differential pressure transducer. By integrating this air flow rate, the tidal volume rate can be determined.

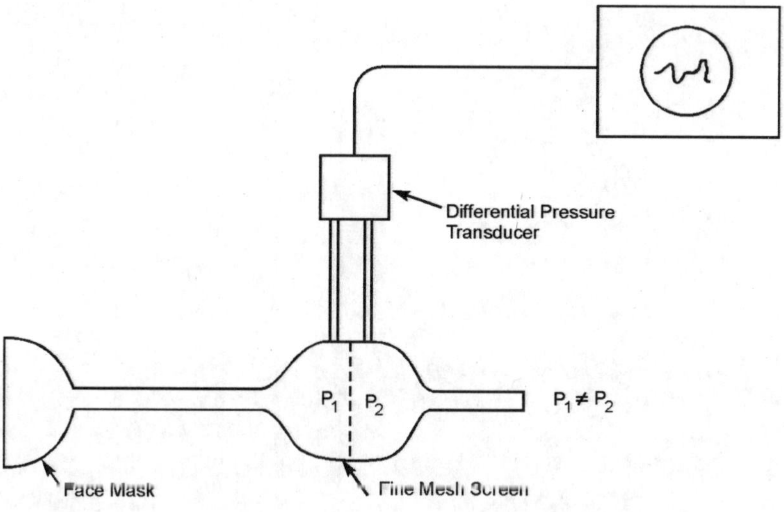

Figure 11.4 *Operating Principle of the Pneumotachograph*

Chapter 12

Special Techniques for the Measurement of Non-electrical Biological Parameters

12.1 With the Biological System as Transducer

Electrical Impedance Plethysmography

Earlier on, in Chapter 5, it has been mentioned that instead of an external transducer, at times the living system itself can be made to modify an externally applied electrical signal. For such measurements, the living system may be termed as an active transducer. A representative example of active transducer measurement is electrical impedance plethysmography.

This technique has been applied to the measurement of the blood content of fingers and limbs, measurement of volume changes of the lungs during breathing and estimation of changes in the stroke volume. The basic principle underlying the plethysmographic measurements is that the impedance of the body segments under study changes during respiration and cardiac activity. Several factors contribute to the impedance changes. Prominent among the factors are the blood-volume changes and changes in shape and alteration in the distribution of blood in the segment. It is difficult to quantitatively determine the contributions of each of the factors individually to the impedance change. But even the estimates possible are useful for biomedical purposes, because plethysmographic techniques do not necessitate any surgical work.

In these techniques, DC is not suitable because it leads to polarization and can be quite painful. Therefore AC, usually in sinusoidal form, is used. Various frequencies have been tried and used. Frequencies below 1 KHz are not suitable because undesirable stimulation of the body segment can occur. As the frequency is increased, the maximum permissible current also increases. For frequencies above 1KHz, currents of the order of a few milliamperes can be used, if the current is not to pass through the heart. If the current passes through the heart, lower current levels are used. When the impedance changes, either the current or voltage changes. If the amplitude of the current is kept constant, the voltage across the body segment will be a modulated sine wave with the envelope corresponding to the impedance changes obtained.

Two types of electrode configurations, known as the *bipolar* and *tetrapolar* systems, are in use. In the bipolar configuration, there are two electrodes. The electrodes are used both for driving in the current as well as for picking up potential changes. The tetrapolar configuration, shown in Figure 12.1, utilizes four electrodes. The two outer electrodes, I_1 and I_2, are used for driving current from a constant current source and the two inner electrodes, E_1 and E_2, are used to pick up potentials. The four electrode configuration has the advantage in that the potential pick-up electrodes do not have to carry a significant current and, therefore, electrode-contact problems are minimized. However, the four electrode configuration has a disadvantage, in that the current paths are not as well defined as in the two electrode system.

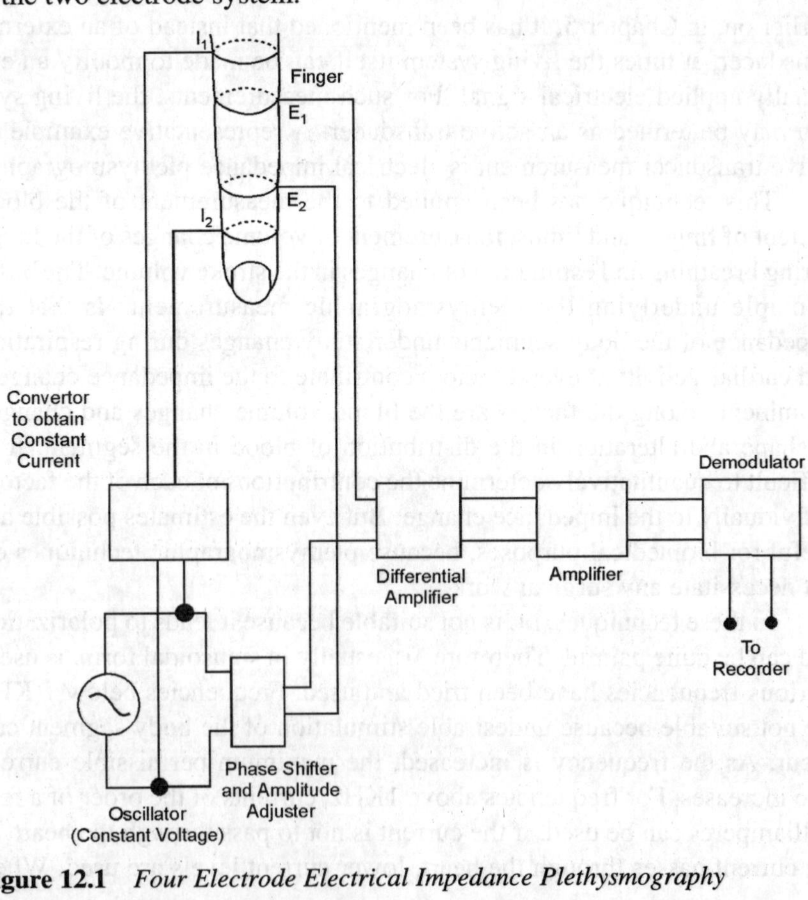

Figure 12.1 *Four Electrode Electrical Impedance Plethysmography*

The current passed through the body segment is kept constant by means of a constant current convertor. For a constant current, I, flowing through the body segment, the potential picked up by the electrodes E_1 and E_2 will be $I(Z + \Delta Z)$, where Z is the impedance and ΔZ the change in impedance. Information regarding $I.\ \Delta Z$ is required. But since $I.\ \Delta Z << I.Z$, before amplification I.Z must be cancelled. For cancellation, the voltage $I(Z + \Delta Z)$ is fed to one pair of terminals of a differential amplifier. To the other ungrounded terminal and the common terminal, a voltage equal to I.Z but opposite in phase is given in such a manner that in the amplifier the two voltages subtract. The cancellation voltage is obtained from a phase shifting and amplitude adjusting device connected as shown in Figure 12.1. After cancellation, the voltage corresponding to $I.\ \Delta Z$ can be amplified and demodulated to give a voltage proportional to ΔZ. The voltage is recorded on a paper recorder. The record is known as a *plethysmogram.*

Figure 12.2 shows a similar configuration for the chest region. In this case, $I.\ \Delta Z$ will have a rapidly varying component in synchrony with the heart pumping action and a slower variation in synchrony with breathing.

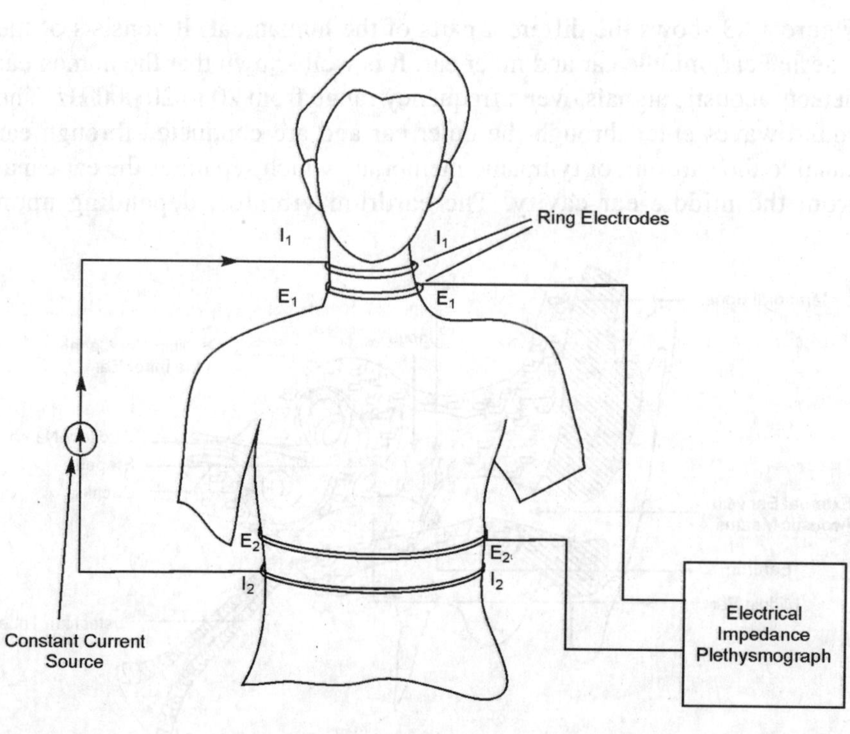

Figure 12.2 *Electrical Impedance Plethysmograph Connected to the Thorax for Monitoring the Heart and Breathing Activity*

12.2 Combined Stimulation and Measurement

12.2.1 Audiometry

The measurement of auditory responses is representative of another class of non-electrical parameter measurements, where stimulation and observation are combined.

The purpose of the measurement of auditory responses is to determine the acuity of hearing, particularly the sensitivity of the ear for various tones. With this technique, the effectiveness of the air and bone conduction mechanisms also can be determined. Air conduction, by definition, is the transmission of sound through the external and middle ear to the internal ear. Bone conduction, on the other hand, refers to the transmission of sound to the internal ear, mediated by the mechanical vibration of the cranial bones and soft tissues.

12.2.2 Anatomy of the Human Ear

Figure 12.3 shows the different parts of the human ear. It consists of the external ear, middle ear and inner ear. It is well known that the human ear detects acoustic signals over a frequency range from 20 to 20 000 Hz. The sound waves enter through the outer ear and are conducted through ear canal to the eardrum or tympanic membrane which separates the ear canal from the middle-ear cavity. The eardrum vibrates, depending upon

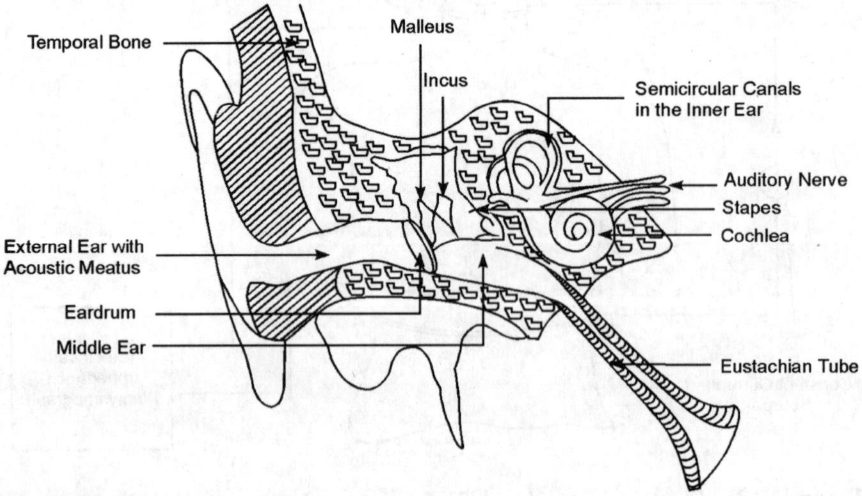

Figure 12.3 *The Human Ear*

the loudness of the sound, and it is critically damped by the auditory ossicles which are connected with the bony wall of the middle ear by two muscles which are responsible for reflex control of hearing sensitivity. From the middle ear, the sound energy is transmitted to the receptor cells in the inner ear which are surrounded by fluid. The ossicular chain, consisting of three small middle-ear bones, functions as acoustic impedance transformer to match air and cochlear fluid mechanical impedance and it improves the efficiency of the sound transmission from the air to the fluid. One of the bones, called the stapes, rests upon the lower end of the cochlea and passes the vibrations directly into the fluid within. The inner ear or cochlea is a fluid-filled, coiled passage in the temporal bone. It is completely divided lengthwise by the basilar membrane containing receptor cells, which transform sound energy into action potentials. The nerve impulses, thus initiated, are propagated along the acoustic (auditory) nerve fibres to the brain at a speed of about 100 m/s. Even though the recognition of notes is partly done by the cochlea, most of the analysis of sound parameters like loudness, pitch and timbre (quality) is done by cerebral activity.

12.2.3 Threshold of Hearing

The threshold pressure level of sound is the lowest level at which an observer can discriminate between the desired sound and the background noise always present in the auditory system. In other words, it is the minimal intensity on the stimulus scale which is just barely adequate to elicit a response. The hearing threshold is not an invariable fixed intensity above which sound is always heard and below which sound is never heard. In fact, the sensitivity of the auditory mechanism is found to vary with interactions between certain physiological, psychological and physical factors. Therefore, the threshold may be regarded as an intensity range within which sound stimuli at or near the statistically determined threshold may or may not be perceived. The base of each audiological examination is the determination of the hearing threshold.

12.2.4 Audiometer

The device used to test auditory responses is the audiometer. Using this electronic–acoustic instrument, the hearing loss is measured in terms of decibels, generally ranging from –10 to 100 dB for the audible range of frequencies. Referring to Figure 12.4, it can be seen that the audiometer is simply a variable frequency sine wave oscillator driving a pair of headphones and calibrated in terms of frequency and acoustic output. The puretone audiometer is widely used to determine hearing loss. Pure tone means a single frequency sound wave. Normally, the pure tone generators generate test tones in octave steps from 125 to 8000 Hz, and the signal intensity can be varied from –10 dB to 100 dB using the attenuator control. The tone generator consists of an LC oscillator in which the inductance and capacitance can be varied precisely and is coupled to a power amplifier. The attenuators used in these instruments are of the ladder-type of a nominal 10 ohm impedance. The signals are delivered to the ear by earphones or headphones. Since the patient response is detected by a signal given by the patient, a patient signalling device in the form of a push button and light is provided.

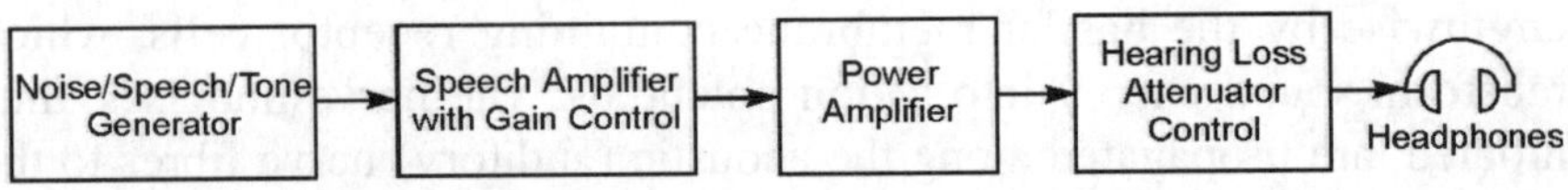

Figure 12.4 *Basic Audiometer*

The simplest form of the audiometer is only suitable for quick and large-scale screening and is not suitable for detailed diagnosis. For accurate diagnosis, additional features must be provided. One of these features is 'masking'. Masking is needed so as to be able to exactly determine the bone conduction capability of each ear individually. With the simple audiometer, the measurement is not possible because one ear is subjected to vibrations for the bone conduction test; the vibrations are carried over to the other ear and the patient or the otologist will not be able to distinguish the response of each ear individually. To obtain masking or suppression of vibrations carried over to the untested ear, an additional masking signal is given to the untested ear. It has been found that effective masking cannot be obtained either by giving a pure harmonic tone to the ear or by giving a white noise which is a combination of all frequencies. Experiments have shown that effective masking is obtained only by giving a white noise filtered so as to contain a narrow band of frequencies around the harmonic tone given to the tested ear. Consequently, a diagnostic audiometer in addition to the harmonic tone oscillator must also have a white-noise generator. Filters must be so incorporated with the white-noise generator that the pass band gets automatically adjusted in synchrony with changes in frequency of the harmonic tone oscillator. The control must be calibrated in decibels.

An additional feature required is the tone interrupter which, when set in operation, will automatically lead to the increase and ultimate decrease of the pure tone oscillator output level.

12.3 Techniques to Aid Observation

There is a class of medical electronic equipment which are not specifically measuring instruments. They only aid in observing some parts of the living system and, therefore, can also be used to measure the size of some organs besides the identification of diseased states. The laryngostroboscope and X-rays are examples of this class of instruments.

12.3.1 Laryngostroboscope

The satisfactory observation of vocal organs, particularly the vocal chords, is not possible in actual operation with ordinary light because the movements are so rapid that the human eye cannot follow the movements. The physical principle of stroboscopy has been used to improve observation.

The basic principle of the stroboscope is that if an object is rotating or oscillating in a repetitive manner rapidly and if bursts of light of short duration illuminate the object, the visual picture differs from the image produced using a steady light. If the repetition rate of light flashing is an exact multiple or submultiple of the cyclic frequency of object movement, the object will appear stationary. The illusion will be most effective if no light other than the flashing light falls on the object and the duration of each flash is so small that the object does not move substantially during that period.

The flash lamp used is a xenon gas-filled lamp. The lamp gives a brief flash of light when an electrical charge passes through it. To enable the charge to flow, an initial high voltage is obtained by stepping up the voltage with a transformer connected to an AC mains supply. The high voltage is converted to DC. By an electronic switching arrangement, working in a repetitive manner, the high voltage is fed to the flash lamps for short durations. For effective observation, the flashing rate must be synchronized with the voice. Therefore, a contact microphone is placed on the head to pick up voice vibrations. The output of the microphone is amplified and fed to a synchronizing circuit. The synchronizing circuit automatically adjusts the repetition rate of the flash-lamp circuit switching, so as to coincide with voice frequency. The phase difference between the light flashing and vocal chord movement can be adjusted by means of a manual potentiometer control.

Some important technical data for a typical laryngostroboscope are:

Source of light	:	Xenon flash lamp with a spectrum similar to that of daylight
Flashing frequency	:	30 to 500 Hz

Continuous power rating of lamp	:	15 VA
Control of lamp	:	(1) Generator range 50–1000 Hz (2) Patient voice in the range 70–900 Hz
Phase variation	:	Full 360°.

12.3.2 X-rays and Radiography

X-rays are electromagnetic radiations with a much shorter wavelength than radio waves or visible light. It is beyond the range of ultraviolet light that the X-rays begin. The shorter the wavelength, the stronger is the penetration power of the X-rays. X-rays normally move in a straight line like light. These rays are used for both diagnostic as well as therapeutic purposes. The use for diagnostic purposes is more prevalent. When they travel through the body of the patient, a portion of them will be either absorbed or scattered, which depends on the density of the medium. It is, therefore, greater in metal than in bone, greater in bone than in soft tissues and greater in soft tissues than in air. The X-rays, thus absorbed, interact with the atoms of the matter and dislodge electrons. This process is called ionization and X-rays are hence referred to as ionizing radiation. The ionization due to the X-rays has a therapeutic effect as well as possible harmful effects. The image of the intensity distribution of the X-rays that have passed through the body of a patient help to visualize the internal structures of the body.

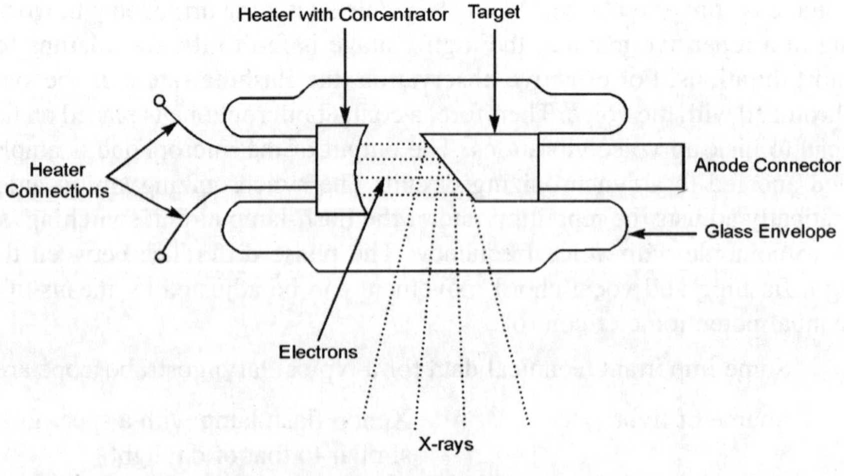

Figure 12.5 *The Principle of X-ray Generation*

X-ray Machine

The main part of an X-ray machine is the X-ray generator. In the X-ray generator, the X-rays are produced in the X-ray tube by bombardment of an anode target with fast moving electrons (Figure 12.5) obtained from a red-hot cathode. The characteristics of the X-ray beam depend on the anode–cathode voltage and the temperature of the electron emitting cathode. The voltages used may vary from 20 KV to several megavolts. The currents are less than a few hundred milliamperes. The X-ray tube for operation requires a high voltage DC source and a low AC voltage source for filament heating. Since control on the high voltage side is not convenient, the controls are incorporated on the low voltage side of the step-up transformer. The filament heating current is obtained from a low voltage AC, obtained by stepping down from the mains supply.

For medical applications, particularly in diagnostic procedures, only short bursts of X-ray are used for photographing images. A longer duration of radiation may be required for screening and therapy. Therefore, a timing unit and switches operated by the timing unit are provided so that the duration of the X-ray beam can be controlled. Figure 12.6 shows schematically the X-ray generating system. Thus, there are three different controls on the control panel of an X-ray machine. The tube voltage expressed in KVP (Kilovolts-peak), determines the hardness or penetration power of the X-ray beam. The beam current, expressed in milliamperes (mA), determines the intensity of the X-ray beam. The third control determines the time (expressed in seconds) that the beam is turned on for X-ray photos.

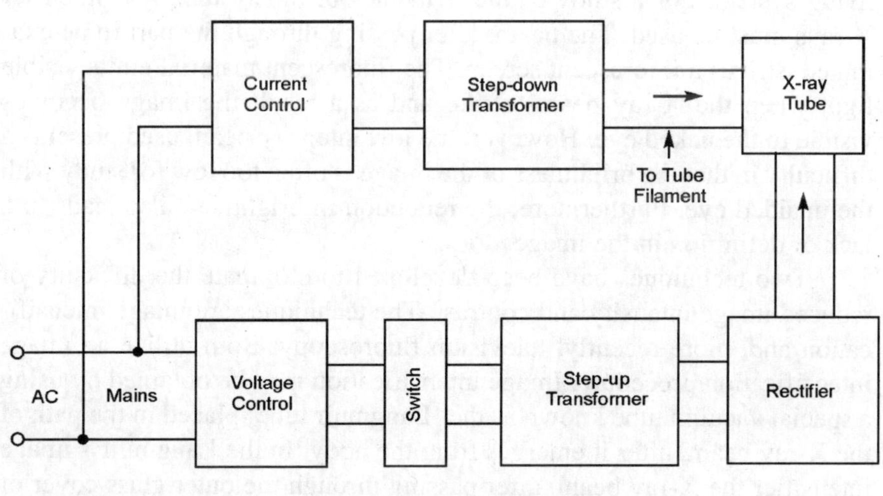

Figure 12.6 *The Block Diagram of an X-ray Generator*

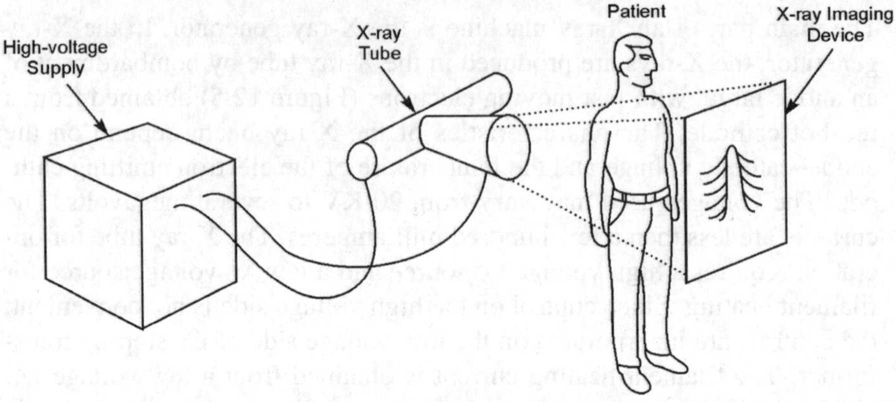

Figure 12.7 *The Use of X-rays to Visualize the Inner Structures of the Body*

As shown in Figure 12.7, the region of the body to be imaged is positioned between the X-ray tube and the imaging device. The X-ray tube projects the "shadow" of the structures inside the body on the imaging device. For the image to be sharp, the part of the body being X-rayed must be as close as possible to the imaging device while the X-ray tube is positioned as far as possible. The X-ray machine can also be mobile, in which case the cassette with the X-ray film is usually placed directly beneath the patient. The X-ray tube is mounted on an arm and can be adjusted to the desired height. The instantaneous picture obtained is useful for a number of diagnostic purposes, but does not allow a study of the dynamics of the living system. For a study of the dynamics of the system, low intensity X-rays must be used. The beam, after passing through the part to be examined, strikes a fluorescent screen. The fluorescent material emits visible light when the X-ray beams strike and as a result the image becomes visible to the naked eye. However, the low intensity beam used presents a difficulty in that the brightness of the image is often too low for study with the unaided eye. Furthermore, the reduction in brightness also leads to a lack of definition in the image.

Two techniques have been developed to eliminate the difficulty of reduced image intensity and contrast. The techniques are image intensification and, more recently, television fluoroscopy. Both utilize an image intensification procedure. Image intensification may be obtained by using a special vacuum tube known as the 'Langmuir tube' placed in the path of the X-ray beam, after it emerges from the body. In the Langmuir's image intensifier the X-ray beam, after passing through the outer glass cover of the intensifier tube, impinges on a fluorescent material. The fluorescent material is backed with a photoelectric film. The light produced by the fluoroscopic material excites the photoelectric film. The electrons emitted are focussed by means of lenses, in a manner similar to the cathode-ray tube. The focussed beam falling on a screen gives an intensified image.

In television fluoroscopy, the contrast is further improved by the scanning of the image. The final image is displayed on a television monitor located at a convenient location. With such a set-up, not only can the image be bright with sharp contrast for low X-ray beam intensities, but in addition, the television monitor can be placed at a convenient location so that the observer is not subjected to hazards of exposure to X-rays.

Contrast Media

X-rays are much more readily absorbed by foreign bodies and bone than soft tissues and visceral organs. It is, therefore, necessary to fill them with a contrast medium before taking the X-ray photo. While taking the pneumoencephalogram, the ventricles of the brain are made visible by filling them with air, which absorbs X-rays less than the surrounding brain structure. In the same way, the structures of the gastrointestinal tract can be made visible with the help of barium sulphate, given orally, which has a higher X-ray absorption than the surrounding tissue. In a similar manner, other body structures and organs can also be visualized by filling them with an appropriate contrast media.

The X-ray techniques described above are two-dimensional presentations of three-dimensional structures. In 'sterioradiography', two X-ray photos are taken from different angles. When viewed in a sterioviewer, a three-dimensional X-ray image is seen. X-ray photos of structures of only a thin section of the body can be visualized by 'Tomography'. Several photos are taken, representing slices at different levels to permit three-dimensional visualization. By moving the X-ray tube and the film cassette in opposite directions during the exposure of the film, tomographic X-ray photos can be obtained. This is called three-dimensional radiography.

X-ray Therapy

Certain diseases and tumors can be treated by the ionizing effect of X-rays. Soft X-rays are used in the treatment of some skin disorders. Deep-seated tumors are treated with very hard X-rays. Concentrated and high energy X-rays are used to destroy cancer cells. Linear accelerators and Betatrons are used to obtain electrons with a very high voltage for this purpose.

Hazards and Safety Precautions

A potential danger to the staff working with X-ray equipment day after day is that they can accumulate dangerous levels of radiation exposure. Some of the hazards are radiation burns, tumors and genetic damage. Radiation badges (containing envelopes with a special film) are to be worn by all personnel working near X-ray equipment. X-ray tubes are normally shielded. The staff member operates the X-ray machine from a small shielded cubicle that has a lead glass window for observation. In fluoroscopy, where the radiologist must attend the patient, aprons and gloves are worn, which are made of plastic containing lead dust.

12.3.3 X-ray Computed Tomography

a. Terminology

X-ray computed tomography is a technology which, when introduced in the '70s, carried the name of 'Computerized Axial Tomography' as well as 'Computer Assisted Tomography'. A more familiar term was CAT Scanning. Other names used were 'Computer Transverse Tomography' and 'Computer Transverse Axial Tomography'. Presently, the commonly used term is CT scanning, which stands for computed tomography scanning.

b. Basic Principles

A problem with X-ray plain film imaging, by the system shown in Figure 12.7, is that the three-dimensional anatomy of the patient is presented as a two-dimensional projection. The gray level of the image on the film at each point represents the sum of X-ray absorptivities of all tissues on the path of the X-ray, from the source to that point on the film. For instance, if a chest radiography is taken, the image of the ribs, lungs and heart overlap in many areas. Also, height and width information with respect to the depth is inadequately displayed. In routine clinical practice, an effort to overcome this limitation is to take a back-to-front view as well as a side-view. Even so, the problem of overlap of body organs in the final image remains. Tomographic imaging attempts to solve the problem by taking a large number of views from different angles. Information from all the data are assembled with the assistance of a computer, so as to obtain an image of one transverse plane of the body with no overlap of images of the different organs. Whereas the conventional X-ray image gives a global view of an entire segment of the body, CT scans give images of a plane. With the recent developments of 'Spiral CT', an assembly of images can be carried out to provide a three-dimensional presentation. Although these differences are present between conventional X-ray and CT, the fundamental fact that the final image is a representation of gray levels proportional to the X-ray absorptivity of tissue remains common to the two methods.

c. *System Hardware*

As the technology has evolved over the past three decades, there have been a number of changes in the system configurations. One feature common in all is that instead of an X-ray film, X-ray detectors which provide an electrical output proportional to the intensity of the X-ray beam falling on the detector have been used. It is the electrical signal from the detectors which are processed for forming the image. The detectors are Xenon detectors or solid state scintillator detectors. The Xenon-type works on the principle of ionization of the gas by X-ray energy. Scintillators which are common in modern CT scanners have some material which 'scintillates', that is, emits light when X-ray energy falls on the material. The emitted light is sensed by a photosensor to provide the electrical output. Both these forms of detectors have a considerably higher sensitivity than the conventional

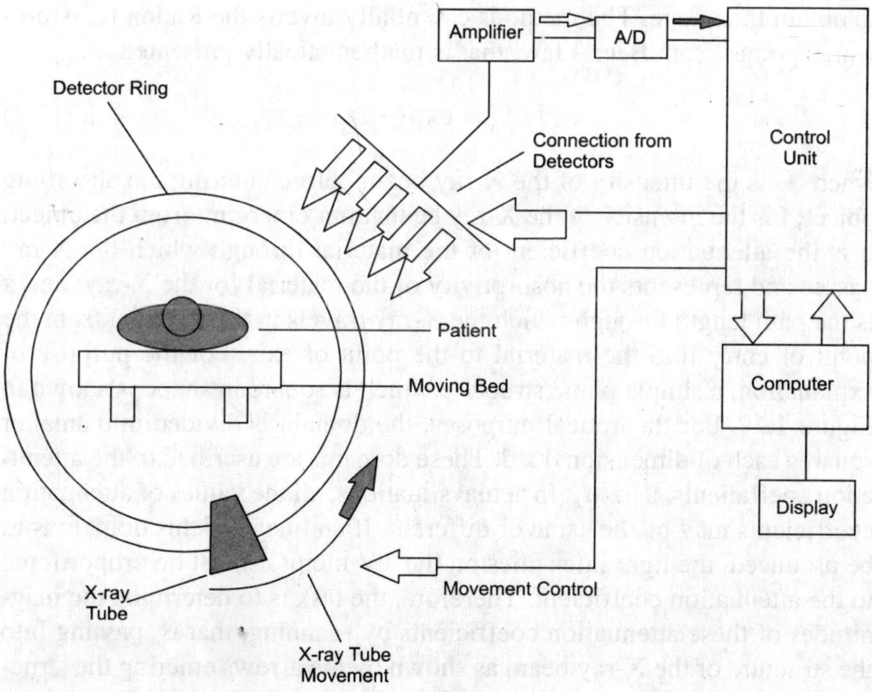

Figure 12.8 *A CT Scan System*

X-ray film. Hence, CT scans may be performed with low intensity X-ray beams and so the X-ray exposure is reduced, thereby improving safety.

As shown in Figure 12.8, the patient lies on a bed. The bed is circled by a 'gantry' which has a ring of detectors. In the figure, an X-ray source is shown to rotate along the gantry. The X-ray detector ring wobbles or mutates to allow the source to pass, maintaining the detectors diametrically opposite in the gantry to the source in the line of the X-ray beam. At each instantaneous position of the source, the electrical output from the detectors on the opposite side of the circle are electronically sampled at a rapid rate, amplified and, after analog to digital conversion by means of analog to digital (A/D) converters, are taken into the computer memory for further processing. Thus, a transverse plane of the body is viewed. In the spiral CT, the bed on which the patient is lying also slowly moves in a direction perpendicular to the plane of the gantry. This act of moving the table continuously, instead of in steps as done earlier, saves scanning time which is now in seconds instead of the earlier minutes.

d. *Image Reconstruction*

There are a number of methods for assembling the data from the detectors to obtain the image. The methods essentially inverts the Radon transform which comes from Beer's law, that is mathematically presented as:

$$I / I_o = \exp(-\mu x) \qquad ....(12.1)$$

where I_o is the intensity of the X-ray beam before entering the absorbing object; I is the intensity of the X-ray beam after emerging from the object; μ is the attenuation coefficient of the material through which the X-ray passes and represents the absorptivity of the material for the X-ray; and x is the path length through which the X-ray travels in the material from the point of entry into the material to the point of exit. For the purpose of explanation, a simple plane structure which is square in shape is shown in Figure 12.9. For theoretical purposes, the domain is divided into smaller squares each of dimension d x d. These domains are ascribed to the attenuation coefficients, $\mu_1 .... \mu_4$. In actual situations, all the values of attenuation coefficients may be the same or different. If an image of this domain is to be produced, the light intensities on the TV monitor must be proportional to the attenuation coefficient. Therefore, the task is to determine the magnitudes of these attenuation coefficients by scanning; that is, passing into the structure of the X-ray beam as shown by the arrows entering the structure and measuring the intensity of the X-ray beam emerging from the structures.

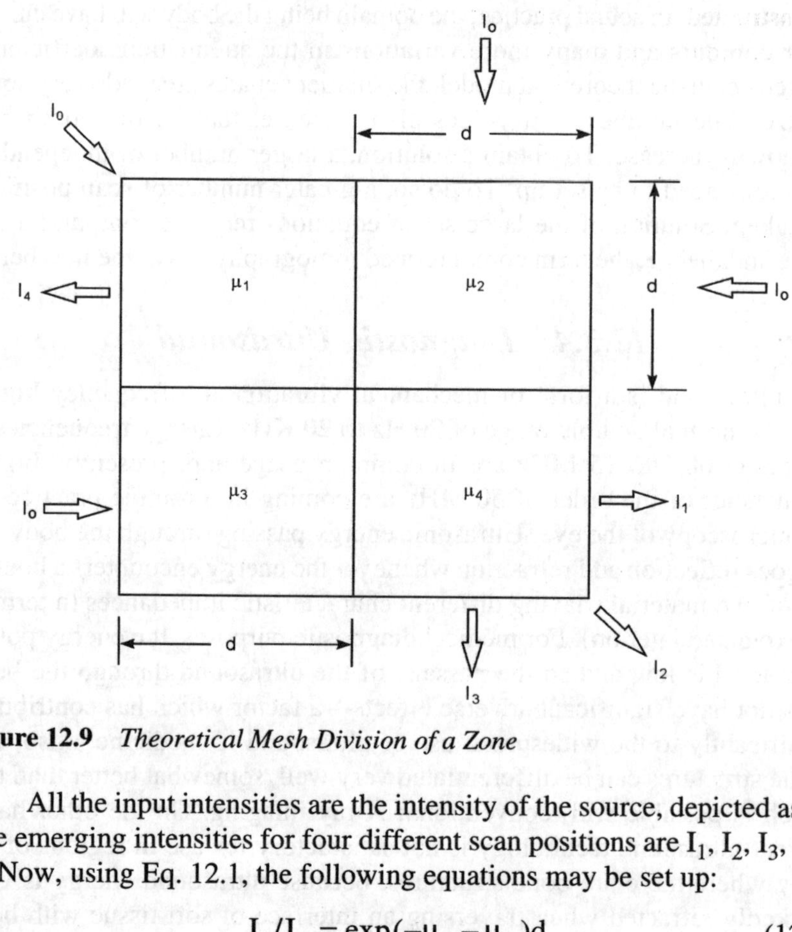

Figure 12.9 *Theoretical Mesh Division of a Zone*

All the input intensities are the intensity of the source, depicted as I_o. The emerging intensities for four different scan positions are I_1, I_2, I_3, and I_4. Now, using Eq. 12.1, the following equations may be set up:

$$I_1/I_o = \exp(-\mu_3 - \mu_4)d \quad ....(12.2)$$

$$I_2/I_o = \exp(-\mu_3 - \mu_4)d\sqrt{2} \quad ....(12.3)$$

$$I_3/I_o = \exp(-\mu_2 - \mu_4)d \quad ....(12.4)$$

$$I_4/I_o = \exp(-\mu_2 - \mu_1)d \quad ....(12.5)$$

In Eq. 12.3, the distances are taken as $d\sqrt{2}$ because the X-ray takes a diagonal path across the squares and by Pythagoras' theorem, the diagonal distance in each small square is $d\sqrt{2}$. Now, the left-hand side of each equation is determined by measurement of the input and exit intensities. So, there are now four equations with the four μ values being the unknowns. This set of equations may be solved by the computer iteration techniques. Thereafter, on obtaining the μ values and giving proportionate intensities on the TV monitor taking care of geometric locations, the image can be reconstructed. In actual practice, the domain being the body will have curved outer contours and many more variations in the attenuation coefficients. Therefore, in the theoretical model, the smaller squares are made very small in size. The number of μ values also increase, that is, the number of unknowns increase. To obtain a solution, a larger number of independent equations need to be set up. To do so, a greater number of scan positions are taken. Solution of the large set of equations requires computer assistance and, hence, the term computerized tomography has come into being.

12.3.4 Diagnostic Ultrasound

The ultrasound is a form of mechanical vibration at a frequency higher than the normal audible range of 20 Hz to 20 KHz. Energy frequencies in the range of 1 to 15 MHz are in common usage and, presently, higher frequencies of the order of 50 MHz are coming into routine practice for biomicroscopy of the eye. Ultrasonic energy passing through the body undergoes reflection and refraction whenever the energy encounters a boundary of two materials having different characteristic impedances (a term to be explained later on). For medical diagnostic purposes, the energy power level used is low and so the passage of the ultrasound through the body does not have significant adverse effects—a factor which has contributed significantly to the widespread use of ultrasound in medicine. Also, soft tissue structures can be differentiated very well, somewhat better than that which is possible with conventional X-ray imaging. On the other hand, current ultrasound technology is not satisfactory for use in regions of the body where there are bone structures, because ultrasound energy is very markedly refracted when traversing an interface of soft tissue with bone and vice versa. Therefore, in ultrasound applications in medicine, the position and orientation of the ultrasound source and the direction of the beam emitted from the source is so selected that the ultrasound passes through an 'ultrasonic window', that is, along a path which does not have bone structures.

a. Physical Principle

In electrical circuits, impedance Z is the ratio of the potential difference and the current; an analogous term in ultrasonics is 'characteristic impedance', which is the ratio of the mechanical particle pressure and particle velocity used. Here, the particles are the small portions of the material through which the ultrasound passes, setting them in a state of vibration. Also, the characteristic impedance is equal to the product of the density of the medium(ρ) and the ultrasonic propogation velocity (v). At an interface of two media, if the characteristic impedance is the same on both sides of the interface, the ultrasonic beam passes across the interface without any reflection. With a mismatch of characteristic impedances at the interface, a part of the energy is transmitted and a part is reflected. For a beam impinging on the interface perpendicular to the surface (Figure 12.10), the ratio of the amplitude of the reflected wave to that of the incident wave (A) is given by

$$A = \frac{Z_2 - Z_1}{Z_2 + Z_1} \quad \text{....(12.6)}$$

where Z_1 is the characteristic impedance of the material in the path of the beam before the interface and Z_2 is the characteristic impedance along the path after the interface. By measuring the amplitude of the incident beam and that of the reflected beam, the characteristic impedance change at the interface can be estimated. For ultrasonic imaging, the change in characteristic

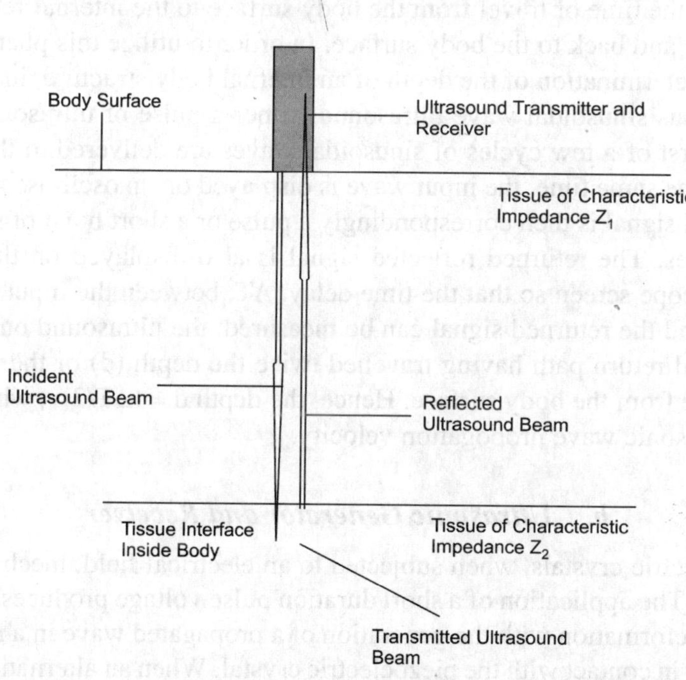

Figure 12.10 *Ultrasound Reflection and Transmission at Tissue Interface*

impedance is represented by changes in gray shade intensity on an oscilloscope screen. Thus, an ultrasonic image is in fact a graphical display of the characteristic impedances of tissues in different regions. For instance, the characteristic impedance of water is 1.48×10^5 g/cm^{2}s and, in an image, a water-filled region is presented as dark. In contrast, the characteristic impedance of fat being lower (1.38×10^5 g/cm^{2}s), a region with fat is depicted in a very light shade.

Ultrasonic technology can also be used to find at what depth from the body surface an organ is situated. When an ultrasonic source is placed on the body surface, that is the skin, and the ultrasonic energy is directed into the interior of the body, there are reflections from the internal organs. Reflections occur because the characteristic impedances of body structures differ. For instance, muscle has a characteristic impedance of 1.70×10^5 g/cm^{2}s, while that of the kidney is 1.62×10^5 g/cm^{2}s. Hence, an ultrasonic beam encountering an interface of muscle and kidney tissue will be partly reflected back towards the source. The reflected component will return to the body surface where it can be sensed by an ultrasonic sensor. Since the waves travel at a finite speed of about 1500 m/s in the body's soft tissue, the reflected wave arrives back at the body surface after a time interval equal to the time of travel from the body surface to the internal reflecting structure and back to the body surface. In order to utilize this phenomena for the determination of the depth of an internal body structure, instead of continuous sinusoidal wave ultrasound, either a pulse of ultrasound or a short burst of a few cycles of sinusoidal waves are delivered to the body and, at the same time, the input wave is displayed on an oscilloscope. The reflected signal is then correspondingly a pulse or a short burst of sinusoidal waves. The returned reflected signal is also displayed on the same oscilloscope screen so that the time delay, ΔT, between the input and the signal and the returned signal can be measured; the ultrasound on its forward and return path having travelled twice the depth (d) of the internal structure from the body surface. Hence, the depth $d = (\Delta T/2)v$, where v is the ultrasonic wave propogation velocity.

b. Ultrasonic Generator and Receiver

Piezoelectric crystals, when subjected to an electrical field, mechanically deform. The application of a short duration pulse voltage produces a short, strong deformation with the generation of a propagated wave in a medium which is in contact with the piezoelectric crystal. When an alternating voltage is applied to the crystal, the expansion and contraction of the crystal along a particular axis takes place and correspondingly, in the contact medium, waves of compression and relaxation are induced which travel in a particular direction much like the travelling wave produced when a pebble is dropped in a pond. The mechanical pulse in the medium for an electrical pulsed input and the mechanical oscillation for an alternating electrical input, have the same frequency spectra as the electrical input.

Piezoelectric crystals, when given a mechanical pulse input, deform. An electrical potential difference is generated across the crystal. The time-course of the electrical signal will be the same as the time-course of the mechanical input. Thus, the piezoelectric crystal can function both as a generator as well as a receiver. The term 'transmitter' is also used synonymously with the term 'generator'. Since ultrasonic crystals convert electrical energy to mechanical energy and vice versa, the term 'transducer' is often ascribed to these devices. Some materials which have the piezoelectric property are quartz, barium titanate, lead zirconium titanate (PZT) and polyvinylidene fluoride (PVDF).

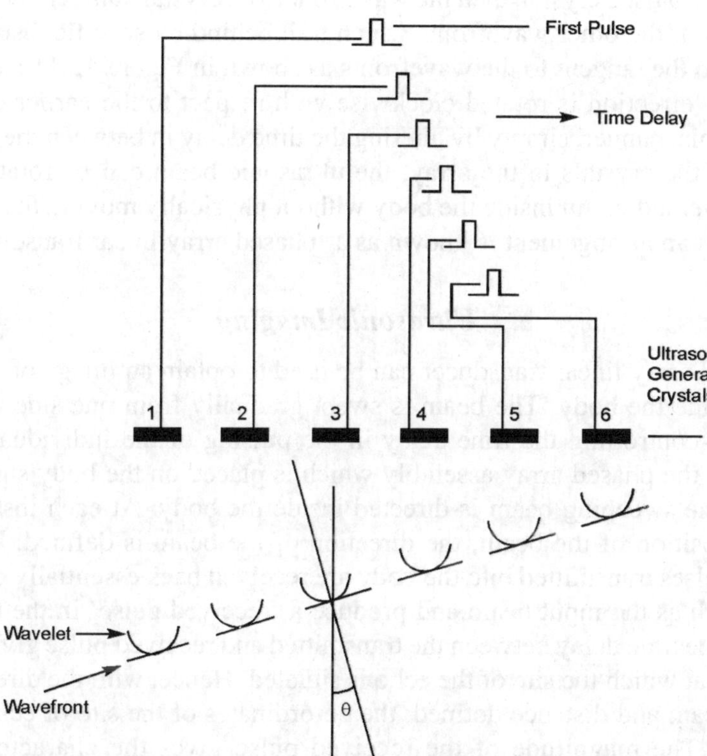

Figure 12.11 *Relation of Electrical Pulse Timing and Direction of Beam Propagation in a Phased Array Transducer*

In ultrasonic imaging, a major application of ultrasound in diagnostics, single piezoelectric crystals do not suffice. From a small transducer assembly placed upon the human body surface, the ultrasonic beam has to be sequentially and rapidly directed to the entire domain to be imaged. The ultrasonic beam itself is narrow and so the beam has to be 'swept' to cover the region of interest. Such a sweeping operation can be performed without moving the transducer, by having an array of piezoelectric crystals as shown in Figure 12.11. In the first case, the crystals are all excited at the same time. Then, at any instant, the ultrasonic wavefronts from all the crystals have travelled an equal distance forward and the net beam direction which is perpendicular to the tangent to the wavefronts will be perpendicular to the line of the crystal array. On the other hand, if crystal number one is pulsed first and then after a specified delay crystal number two is pulsed and succeeding crystals are pulsed each after the same delay to the previously pulsed crystal, then the wavefront from crystal number one travels ahead of the other wavefronts which trail behind by specific distances leading to the tangent to the wavefronts as shown in Figure 12.11 and the net beam direction is rotated clockwise with respect to the earlier direction. In this manner, simply by altering the time delay in between the excitation of the crystals in the array, the ultrasonic beam can be rotated to sweep over a domain inside the body without physically moving the crystals. Such an arrangement is known as a 'phased array linear transducer'.

c. *Ultrasonic Imaging*

A phased array linear transducer can be used to obtain an image of structures inside the body. The beam is swept gradually from one side to the other by controlling the time delay in the pulsing of the individual elements of the phased array assembly which is placed on the body surface, so that the sweeping beam is directed inside the body. At each instantaneous position of the beam, the direction of the beam is defined. Ultrasound pulses transmitted into the body are received back essentially on the same path as the input beam and produce a 'received pulse' in the transducer. The time delay between the transmitted and received pulse gives the distance at which the site of the echo is situated. Hence, with the direction of the beam and distance defined, the co-ordinates of the site of echo are defined. The magnitude of the received pulse gives the characteristic impedance change at the echo region. All the data is processed in a 'digital scan converter', schematically shown in Figure 12.12. The echo signals are in analog form and, by means of an A/D converter, the digital values are taken into the memory following a stage of 'pre-processing'. The spot on a TV monitor is positioned appropriately according to the determined co-ordinates of the site of the echo. The intensity of the gray level of the spot is proportional to the magnitude of the echo signal. This procedure is

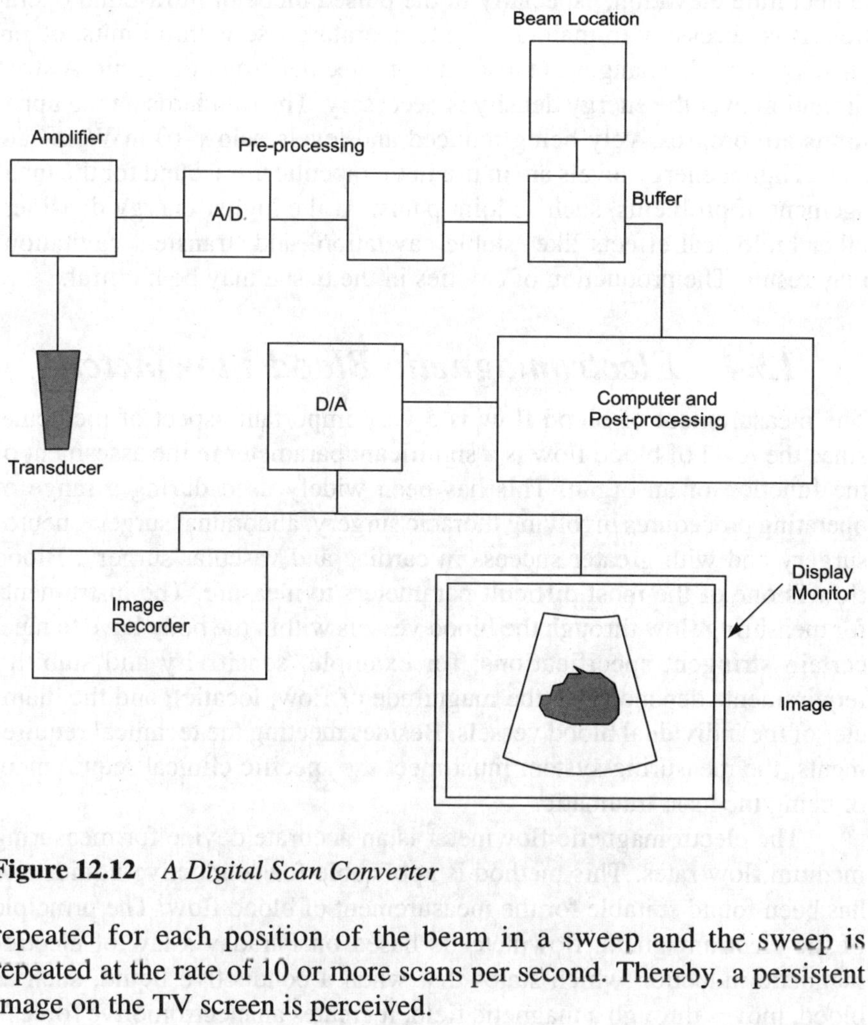

Figure 12.12 *A Digital Scan Converter*

repeated for each position of the beam in a sweep and the sweep is repeated at the rate of 10 or more scans per second. Thereby, a persistent image on the TV screen is perceived.

d. *Ultrasonic Safety and Standardization*

Ultrasound, at the energy levels prevalent in diagnostic medicine, is generally considered 'safe'. However, if the equipment is not designed and calibrated, a higher energy level may be delivered, leading to adverse effects. Since ultrasonic imaging is routinely performed in the field of obstetrics on pregnant mothers and the fetus is quite susceptible to harm, ultrasonic safety and standardization are important considerations.

Ultrasound, even at low levels, has a thermal effect and raises the temperature of the tissue. The tissue has fairly high absorptivity and low thermal conductivity and these two factors together promote localized temperature elevation, especially in the pulsed mode of ultrasound operation. It is necessary to maintain the temperature rise within limits, otherwise irreversible changes in the tissue's molecular structure occur. A strict limitation over the energy density is necessary. The standards on the upper limits are progressively being reduced and levels below 10 mW/cm^2 are used. Higher energy levels are in use in therapeutic ultrasound for the management of problems, such as joint pains. At the higher energy densities, other biological effects like 'stable cavitation' and 'transient cavitation' may result. The production of cavities in the tissue may be harmful.

12.4 Electromagnetic Blood Flowmeter

The measurement of blood flow is a very important aspect of medicine, since the level of blood flow is a significant parameter in the assesment of the function of an organ. This has been widely used during a range of operating procedures involving thoracic surgery, abdominal surgery, neurosurgery and with greater success in cardiac and vascular surgery. Blood flow is one of the most difficult parameters to measure. The instruments for measuring flow through the blood vessels within the body have to meet certain stringent specifications; for example, sensitivity and stability requirements depend upon the magnitude of flow, location and the diameter of the individual blood vessels. Besides meeting the technical requirements, the measuring system must meet the specific clinical requirement of being the least traumatic.

The electromagnetic flowmeter is an accurate device for measuring medium flow rates. This method is applicable for conductive liquids and has been found suitable for the measurement of blood flow. The principle of the electromagnetic flowmeter is based on Faraday's law of electromagnetic induction which states that when a conductive liquid, such as blood, moves through a magnetic field, it creates an electromotive force V along an axis at right angles to the magnetic field and the direction of motion. If the flow is laminar and the tubing is circular with a fixed diameter, the emf V_f produced is directly proportional to the flow rate. It would, therefore, be quite simple to have a rigid tubing with two electrodes to pick up the emf and place the tube between the jaws of a horse-shoe magnet. Blood or any other liquid could be made to flow into the tubing and back into the living system with minimal disturbance to the biological system. The emf produced, which is of the order of 10 μV, can be amplified and recorded to give the instantaneous flow rate. But blood, being ionic, tends to polarize at the electrodes, disrupting the measurement process. To avoid polarization, about a dozen different types of electromagnetic flowmeters have been designed, based on the principle stated. A simple and effective type is the *sine-wave flowmeter.*

In the sine-wave flowmeter, the magnet is an electromagnet excited with a sinusoidal alternating current at 100 Hz. If the magnet is energized by the sinusoidal AC voltage, the electrodes and their leads act as a one-turn transformer coil in which an additional transformer emf V_t is induced. The electrodes, therefore, sense a voltage $V = V_f + V_t$, where both V_f and V_t are sinusoidal. V_t is in phase quadrature with respect to V_f in time-phase. The amplitude of V_f is proportional to the flow rate and V_t contains no useful information. By bringing out the leads from the electrodes, so that the plane of the loop is parallel to the magnetic lines of flux, V_t is minimized. The orientation is shown in Figure 12.13.

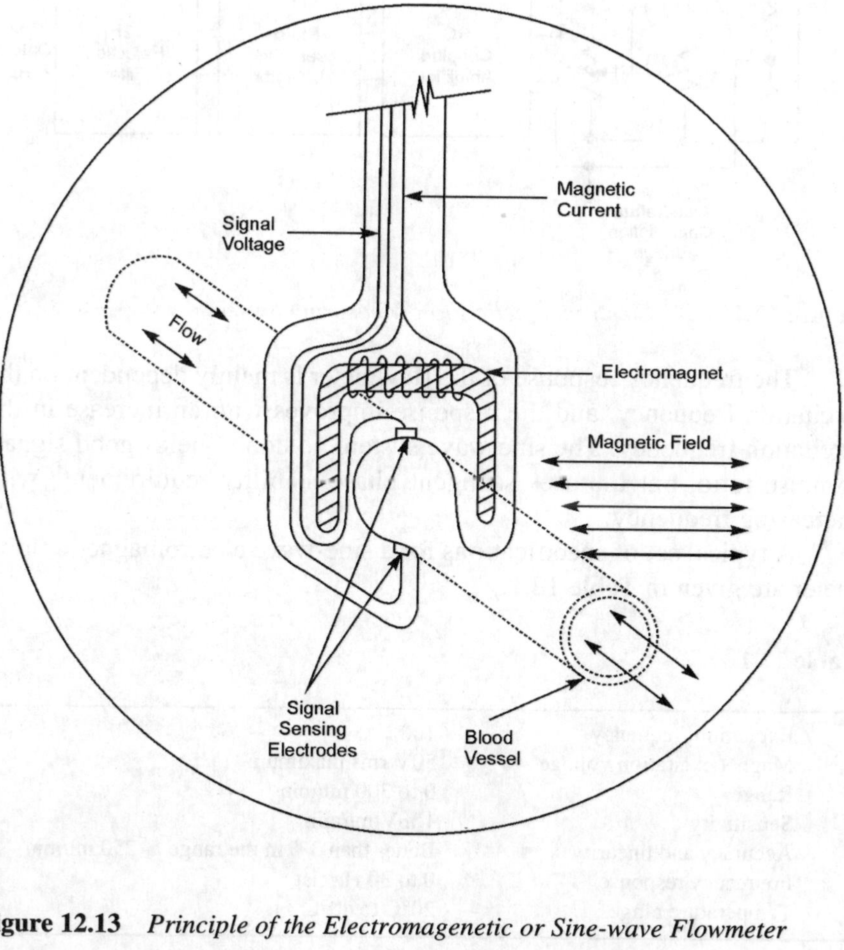

Figure 12.13 *Principle of the Electromagenetic or Sine-wave Flowmeter*

Further reduction of the quadrature component is obtained by a quadrature cancellation bridge, as shown in Figure 12.14. After cancellation, the emf is amplified and demodulated by means of a phase-sensitive detector, which rejects any remaining portion of V_t and extracts the envelope of V_f. The output of the demodulator, after filtering, is fed to a recorder.

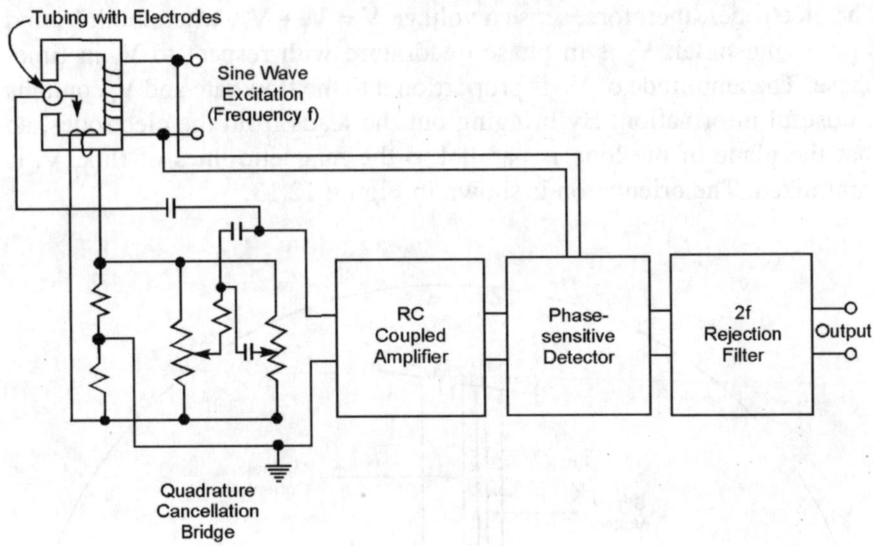

Figure 12.14 *An Electromagnetic Flow Measuring System*

The frequency response of the flowmeter is mainly dependent on the excitation frequency, and the response improves with an increase in the excitation frequency. The sine wave system no doubt yields good signal-to-noise ratio, but imposes stringent phase stability requirements with increasing frequency.

A typical set of specifications for a sine-wave electromagnetic flowmeter are given in Table 12.1.

Table 12.1

Excitation frequency	100 Hz
Magnet excitation voltage	50V rms maximum
Range	0 to 300 ml/min
Sensitivity	1 mV/ml/min
Accuracy and linearity	Better than 3% in the range 5–250 ml/min
Frequency response	0 to 30 Hz flat
Temperature range	20°C to 40°C

Chapter 13

Electronic Instruments for Affecting the Human Body

13.1 Stimulators

The bioelectric potentials generated within individual cells, when these cells are stimulated, were discussed in Chapter 2. Stimulation refers to an external influence, generally electrical, mechanical or chemical, being applied to the cell which results in the cell depolarizing and then repolarizing. All cells are sensitive, to some degree, to artificial electrical stimulation. When analyzing biomedical phenomena, it is often desirable to artificially stimulate a group of cells. Such artificial stimulation is achieved by using a pulse generator to pass a current through the cells concerned for a brief period of time.

13.1.1 Strength/Duration Curves

Strength/duration curves show the excitability characteristics of muscle or nerve fibres. Stimulation of a cell is achieved by the passage of a certain quantity of electricity through the cell; thus, stimulation is dependent on charge rather than on current. Stimulation can be achieved either by passing a large current for a short period or a lower current for a longer period of time. Since diffusion within the cell tends to oppose the stimulating current, there is a lower limit of current below which stimulation will not occur, no matter how long the current is maintained.

For any given muscle or nerve fibre, a strength/duration curve may be drawn, as shown in Figure 13.1, which represents the minimum stimulus or threshold required to stimulate the muscle or nerve fibre. Referring to the strength/duration curves shown in Figure 13.1, for a normal muscle it can be seen that a 0.01 m/s stimulus pulse of 3.5 amplitude units (volts) applied in the region of the nerve associated with the muscle will cause the nerve to depolarize, causing muscular action. Alternatively, muscle action can be produced with a 0.03 m/s-wide stimulus pulse of 1.7 amplitude units, a 0.1 m/s-wide stimulus pulse of 1.1 amplitude units or a 0.3 m/s-wide stimulus pulse of 1.0 amplitude units. It is also apparent that, for pulse widths greater than 0.03 m/s, no further reduction in stimulus amplitude will be effective. This amplitude, below which stimulation will not occur, no matter what the stimulus width, is referred to as the nerve tissue threshold or the *rheobase* for the muscle. The term is now rarely used. Pulse widths from 0.03 m/s to 100 m/s in eight steps have been proposed as an international standard for strength/duration curve determination.

In the particular example shown in Figure 13.1, this rheobase occurred when using a constant voltage stimulator set at 30 V. Since, as discussed earlier, this absolute voltage level has little meaning, strength/duration curves are normally shown with a stimulus amplitude normalized to the rheobase, that is 30 V, which, in this instance, is equivalent to 1.0 unit of stimulus amplitude; 2.0 units of stimulus amplitude being equivalent to 60 V.

Once a strength/duration curve has been determined, it is desirable to have a technique whereby curves for various muscles can be compared. Strength/duration curves may be compared by comparing the chronaxy obtained from the curve. The chronaxy is the minimum pulse width required to excite the tissue for a stimulus of twice the amplitude of the rheobase. The chronaxy, in the example shown in Figure 13.1, is 0.022 m/s for a normal muscle.

The strength/duration curves for a denervated muscle is also shown in Figure 13.1. A denervated muscle exists when the nerve connection to the

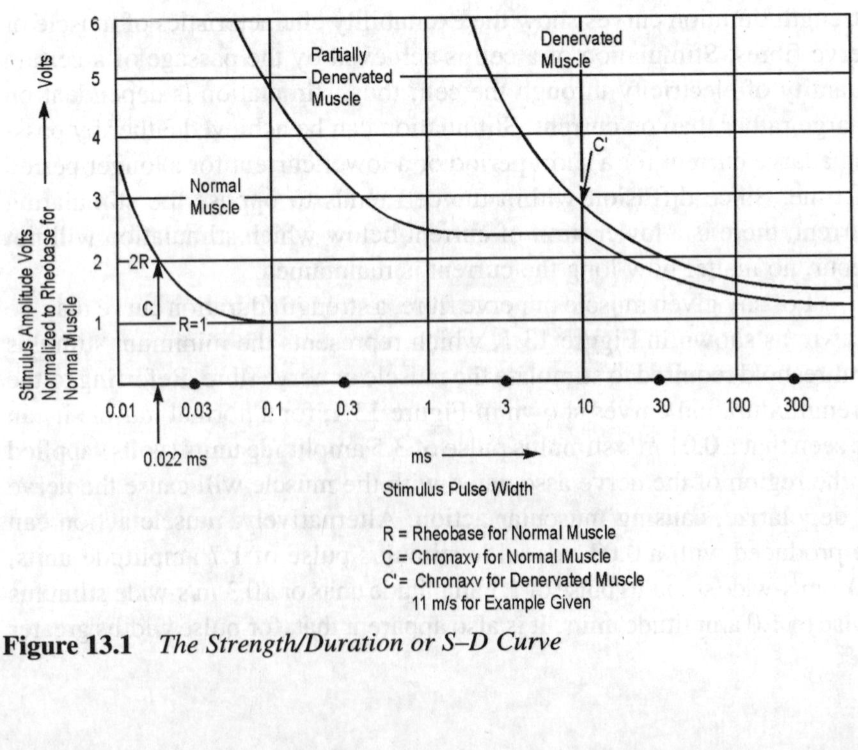

Figure 13.1 *The Strength/Duration or S–D Curve*

muscle has effectively been interrupted. This may be achieved by impairing the motor end-plate action with drugs such as curare. It can be seen from Figure 13.1 that a denervated muscle is less sensitive to stimulus than a normal muscle. In a normal muscle, the stimulating current excites the more sensitive nerve fibres which, in turn, excite the muscle; whereas, in a denervated muscle, the stimulating current must stimulate the muscle directly. Thus, the S–D curve for a normal muscle shows the excitability characteristics for a single motor neuron. In contrast, however, the S–D curve for a denervated muscle shows the stability characteristics for a single muscle fibre.

Figure 13.1 also shows a S–D curve for a partially denervated muscle. It can be seen that for pulse widths up to 10 m/s, muscle action is due to stimulation of the nerve; whereas, for pulse widths greater than 10 m/s, muscle action is due to stimulation of the muscle directly. The S–D curve is thus displaying the excitability characteristics of the component which has a lower threshold for a specific pulse width.

In some instances, the plotting of the S–D curve is somewhat tedious and abbreviated S–D curves are obtained by determining the stimulus amplitude required for a 100 m/s and a 1 m/s pulse and expressing these as a ratio. In a normal muscle this ratio must be approximately unity, in a partially denervated muscle it will be between 1.5 and 4 and in a fully denervated muscle it will be between 4 and infinity. Abbreviated S–D curves may also be obtained by determining the rheobase and then determining the chronaxy directly by doubling the stimulus amplitude and reducing the pulse width to a point where the tissue can no longer be stimulated. This point will represent the chronaxy.

13.1.2 Types of Stimulators

Stimulators are the devices which are used to stimulate innervated muscles, denervated muscles and nerves. Further, these are used for the treatment of paralysis with totally or partially denervated muscles, for the treatment of pain, muscular spasms and peripheral circulatory disturbances. This technique is called 'Electrotherapy' which uses low voltage, low frequency impulse currents.

The physiological action of stimulation depends on the passage of current across the cell membrane. Therefore, it may be expected that the magnitude of the current (as well as its waveform) is an important factor. It is often overlooked that for stimulation, the applied voltage and current are two distinct parameters. They are, no doubt, related and the relationship depends on the source and load impedance.

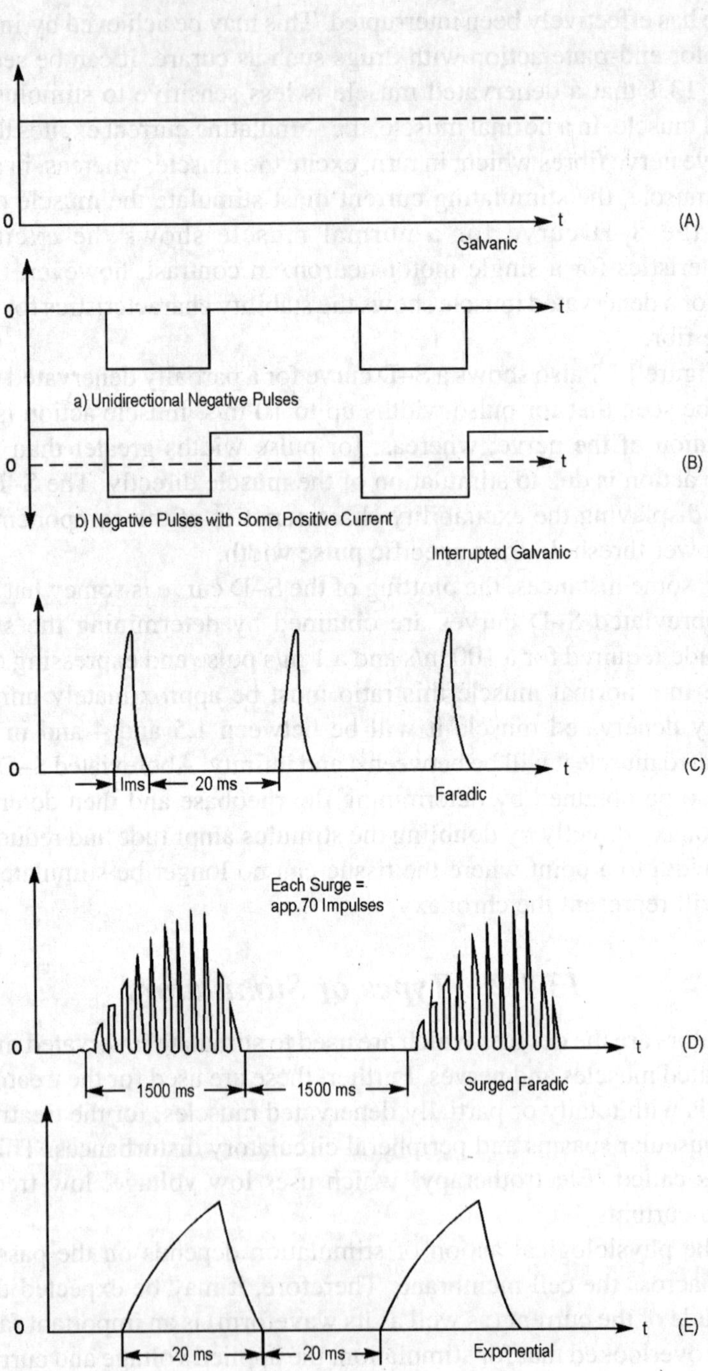

Figure 13.2 *Different Types of Stimulator Waveforms*

Different types of waveforms are used for carrying out treatments of a specific nature. The most commonly used pulse waveforms are described below.

a. Galvanic Current

When a steady flow of direct current, see Figure 13.2(A), is passed through a tissue, its effect is primarily chemical. It causes the movement of ions and their collection at the areas of the skin immediately beneath the electrodes. The effect is manifested most clearly in a bright-red colouration which is an expression of *hyperaenia* (increased blood flow). Galvanic current is also called direct current, galvanism or continuous current.

Galvanic current may be used for the preliminary treatment of atonic paralysis and for the treatment of disturbances in blood flow. It is also used for iontophoresis, which is the introduction of drugs into the body through the skin by electrolytic means. In general, the intensity of the current passed through any part of the body does not exceed 0.3 to 0.5 mA/cm^2 of electrode surface. The treatment duration is generally 10–20 min.

b. Interrupted Galvanic Current

In an interrupted galvanic current, pulses are a series of negative, ongoing, rectangular pulses. The pulse duration is about 100 m/s with a repetition rate between 12/min and 70/min. Figure 13.2(B)(a) shows the unidirectional, interrupted galvanic pulses which create the ionization of the patient's skin and produces discomfort and inflammation. The discomfort is reduced by the application of a positive current, Figure 13.2(B)(b), in between the negative pulses and proportional to the time interval. That is, a charge balance is obtained.

c. Faradic Current

Faradic current is a sequence of pulses with a defined shape and current intensity (Figure 13.2(C)). The pulse duration is about 1 m/s with a triangular waveform and an interval duration of about 20 m/s. Faradic current acts upon muscle tissue and upon the motor nerves to produce muscle contractions. There is no ion transfer and, consequently, minimal chemical effects. Such stimuli may be used for treatment of muscle weakness after lengthy immobilization and of disuse atrophy.

d. Surged Faradic Current

If the peak current intensity applied to the patient increases and decreases rhythmically, Figure 13.2(D), and the rate of increase and decrease of peak amplitude is slow, the resulting shape of the current waveform is called a surging current.

The main field of application of Faradic surge current is in the treatment of functional paralysis. The surge rate is usually from 6–60 surges per minute in most of the instruments. The ratio of the interval to the duration of the surging is also adjustable, so that a graded exercise may be administered. Each surge has a duration of 1500 m/s and approximately 70 impulses. Its repetition rate is about 2 to 3 s. This type of current is usually required for the treatment of spasms and pain.

e. *Exponential Current*

Figure 13.2(E) shows the exponential pulse used for the treatment of severe paralysis. With these kind of pulses, the surrounding healthy muscles, even in the immediate neighbourhood of the diseased muscles, are not stimulated. The slope of the exponential pulse can be varied to provide selective stimulation.

13.1.3 An Electrodiagnostic/ Therapeutic Stimulator

Several types of commercial units are available, which give specific output waveforms for specific applications. However, the trend is in favour of having a versatile apparatus which gives output current waveforms to cover the whole range of electrodiagnostic and therapeutic possibilities. Another important consideration is that the apparatus must be either of constant voltage or constant current type. Moreover, the apparatus must give reproducible and well-defined impulses that must correspond to the value set on the dials. For clinical practice, the maximum tolerance emitted in the pulse parameters is ±5%. The instrument generally has a floating output and incorporates an isolation transformer in the output.

13.1.3.1 Typical Specifications of Electrodiagnostic/ Therapy Unit

- Galvanic current upto 80 mA; with ripples less than 0.5% as a constant current or surging current with adjustable surge frequency from 6 to 30 surges per minute.
- Exponentially progressive current pulse sequences with continuously variable pulse duration from 0.01 to 1000 m/s and independently adjustable interval duration of 1 to 10 000 m/s. The pulse form can be set continuously between triangular and rectangular.

- Faradic surging current with 25 surges per minute, up to 80 mA. Precision and constancy of the values set better than ±10%; peak current measurement facility. Constant current circuit—both poles are earth-free.

13.1.3.2 Functional Schematic Diagram — A Description

Figure 13.3 shows the block diagram of a versatile electrodiagnostic therapeutic stimulator. It makes use of a variable multivibrator (M_1) to set the basic stimulus frequency. The output from a free-running multivibrator triggers a monostable multivibrator (M_2) circuit which sets the pulse width. The output pulse from the M_2 provides an interrupted galvanic output, whose rate as well as duration can be independently controlled. Another astable multivibrator produces short duration pulses called *Faradic currents.* Faradic currents are usually modulated at the frequency set by M_1 in a mixer circuit (M_4). Since the modulation of Faradic pulses takes place with a slow rate of increase and decrease, the output of M_4 is surged Faradic current. By integrating the output of M_2, the interrupted galvanic pulses can be modified to have an exponential rise and fall. The shape of these pulses is similar to the triangular waveform. Galvanic current is also made available by suitably tapping the DC supply.

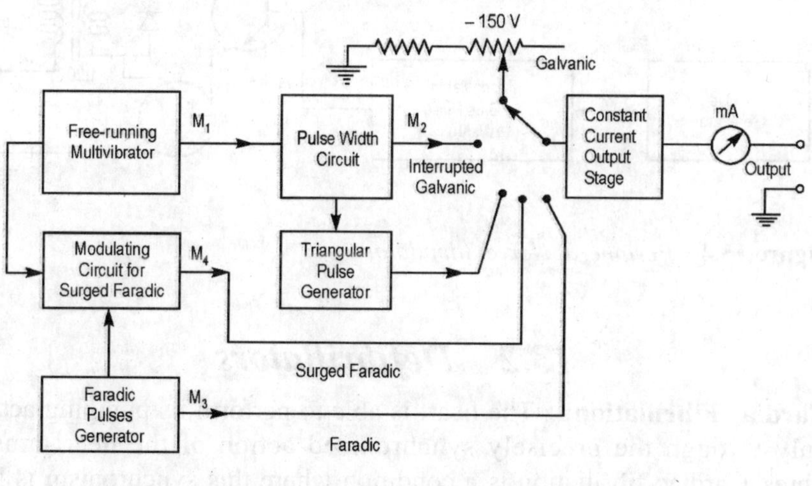

Figure 13.3 *Block Diagram of a Versatile Electrodiagnostic/ Therapeutic Stimulator*

Any one of the above cited waveforms can be selected through a selector switch and fed to an emitter–follower input, to either get a low output impedance constant voltage output or to get a high output impedance constant current stage. Usually, the output impedance of a constant voltage stimulator is of the order of 100 ohms and of a constant current type being greater than 100 Kohms. The output of a constant voltage stimulator is indicated on a meter calibrated in volts, whereas in a constant current stimulator, the meter is calibrated to read the current directly in milliamperes in the patient circuit. The output of this unit is kept floating (or) isolated from the earth. There is an isolation transformer at the output of the stimulator.

13.1.4 Peripheral Nerve Stimulator

Figure 13.4 shows the circuit for a peripheral nerve stimulator. The pulse generator (astable multivibrator) which determines the pulse repetition rate generates repetitive negative rectangular pulses, which trigger the monostable vibrator, which determines the pulse width. The output of the monostable vibrator drives an emitter–follower and a transconductance amplifier which increases the maximum output voltage. Finally, the transformer is used to couple the stimulator and the skin with suitable energy to stimulate the nerve trunk.

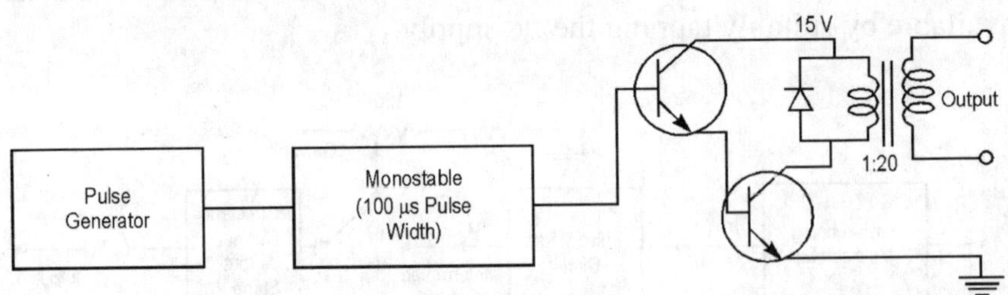

Figure 13.4 *Peripheral Nerve Stimulator*

13.2 Defibrillators

Cardiac Fibrillation: The heart is able to perform its pumping action only through the precisely synchronized action of the heart-muscle fibres. Cardiac fibrillation is a condition where this synchronism is lost, the individual myocardial cells contract asynchronously and in a rapid, irregular cardiac rhythm. It is one of the most serious medical emergencies of the cardiac patient, as the condition reduces the cardiac output to near zero. The fibrillation of atrial muscles is called atrial fibrillation and that of ventricles is called ventricular fibrillation.

13.2.1 Mechanism

Fibrillation results from a rapid discharge of impulses from single or multiple foci in the atria or in the ventricles. The atria or the ventricles are unable to respond completely and effectively to each stimulus.

Under conditions of atrial fibrillation, the ventricles can still function normally, but they respond with irregular rhythms to the non-synchronized bombardment of electrical stimulation from the fibrillating atria and the circulation is still maintained, although not as efficiently.

The sensation produced, however, by the fibrillating atria and irregular ventricular action can be quite traumatic for the patient. Ventricular fibrillation is far more dangerous, for under this condition the ventricles are unable to pump blood. Hence, resuscitative measures must be applied within 5 minutes or less after the attack, or irreversible brain damage and death will occur.

13.2.2 AC Defibrillators

Although mechanical methods like heart massage for defibrillation have been tried for years, the most successful method of defibrillation is the application of electric shock to the area of the heart which makes all the heart muscle fibres enter their refractory periods together, after which normal heart action may resume. One of the earliest forms of an electrical defibrillator was the AC defibrillator, which applies several cycles of alternating current to the heart from the power line through a step-up transformer. To achieve defibrillation with internal electrodes placed on the surface of the heart (in open heart surgery), voltage ranging from 80 to 300 V rms is required. When external electrodes are used on the chest, voltages of twice this value are required. The transformer must be capable of supplying 4 to 6 amperes during the stimulus period.

There are a number of disadvantages in using AC defibrillation. Successive attempts to correct ventricular fibrillation are often required. Further, AC defibrillation cannot be successfully used to correct atrial fibrillation. DC defibrillators have, therefore, now replaced AC defibrillators. The DC signal overcomes the above stated shortcomings of AC defibrillators. In addition, the DC signal has fewer deleterious effects on the heart than the AC pulse and is less likely to produce ventricular fibrillation when applied at random during the cardiac cycle.

13.2.3 Capacitive Discharge DC Defibrillators

The DC defibrillator is of the capacitive discharge type, the circuit of which is shown in Figure 13.5(a). The 220 V AC mains supply is connected to a variable autotransformer in the primary circuit. The output of this is fed to a step-up transformer to produce high voltage with a rms value of about 8 000 V. A half-wave rectifier rectifies this high AC voltage to obtain high

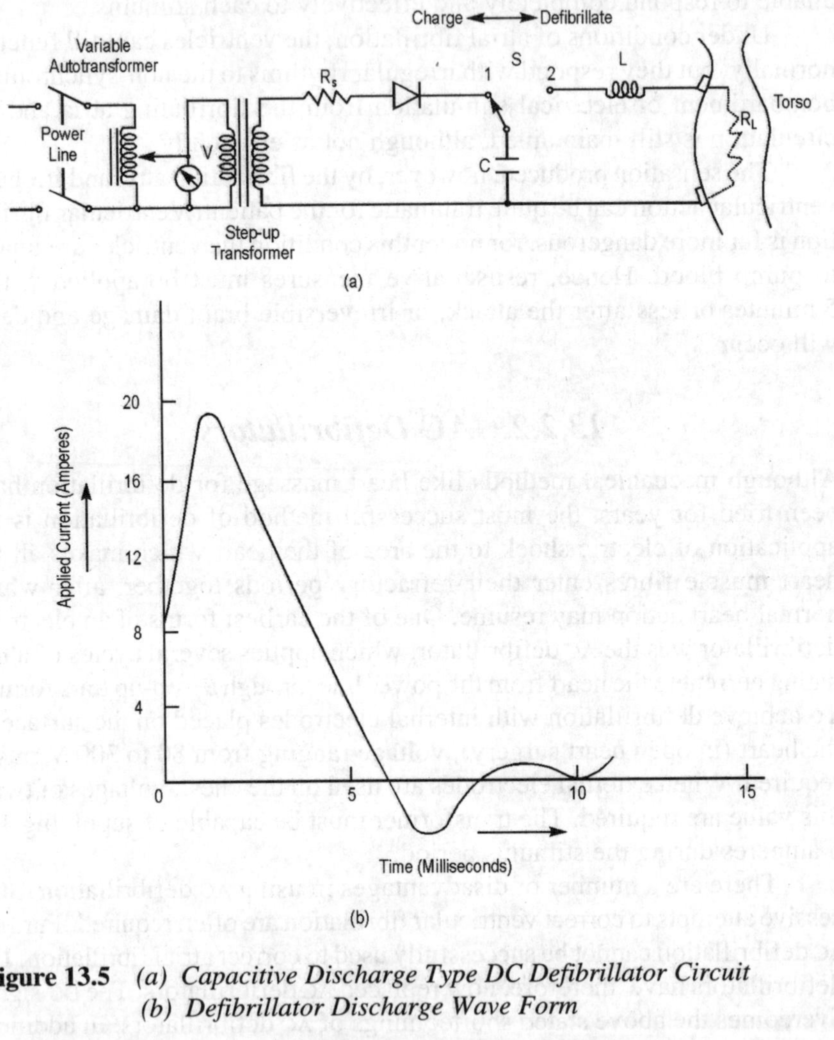

Figure 13.5 *(a) Capacitive Discharge Type DC Defibrillator Circuit*
(b) Defibrillator Discharge Wave Form

DC voltage, which charges the capacitor C. The voltage to which C is charged is determined by the autotransformer in the primary circuit. A series resistance, R_S, limits the charging current to protect the circuit components and an AC voltmeter across the primary is calibrated to indicate the energy stored in the capacitor. Five times the RC time constant for the circuit is required to reach 99% of a full charge—a value it should reach in 10 s, which means that the time constant must be less than 2 s.

With the electrodes firmly placed at appropriate positions on the chest, the clinician discharges the capacitor by momentarily changing the switch S from position 1 to position 2. The capacitor is discharged through the electrodes and the patient's torso represented by a resistive load, and the inductor L. The inductor is used to shape the wave in order to eliminate a sharp, undesirable current spike that would otherwise occur at the beginning of the discharge. The energy delivered is represented by the typical waveform shown in Figure 13.5(b). The area under the curve is proportional to the energy delivered. The wave is essentially monophasic and the peak value of the current is nearly 20 A. Depending on the defibrillator energy setting, the amount of electrical energy discharged by the capacitor may range between 100 and 400 watt-s, or joules when the electrodes are applied externally and the duration of the effective portion of the discharge is approximately 5 m/s. Once the discharge is completed, the switch automatically returns to position I and the process can be repeated, if necessary. When the electrodes are applied directly to the heart, about 50 to 100 joules only is required for defibrillation. The energy stored in the capacitor is given by the equation

$$W = \frac{1}{2} CV^2 \qquad \text{....(13.1)}$$

where C is the capacitance and V is the voltage to which the capacitor is charged. Capacitors used in the defibrillator range from 10 to 50μF. Thus, the voltage for a maximum of 400 J ranges from 2 to 9 kV, depending on the size of the capacitor. Losses in the discharge circuit and at the electrodes result in lower actual delivered energy.

13.2.4 Delay-Line Capacitive Discharge DC Defibrillator

Even with DC defibrillation, there is a danger of damage to the myocardium and the chest walls, because peak voltages as high as 6 000 V may be used. To reduce this risk, some defibrillators produce dual-peak waveforms of longer duration (approximately 10 m/s) at a much lower voltage. The circuit diagram of such a system is shown in Figure 13.6(a). The parallel combination of C_1 and C_2 stores the same energy as the single capacitor in Figure 13.5(a). But its discharge characteristic is more rectangular in shape (longer duration of approximately 10 m/s) at a much lower voltage, as shown in Figure 13.6(b). With this type of waveform, effective defibrillation can be achieved in adults with lower levels of delivered energy—between 50 and 200 watt-s.

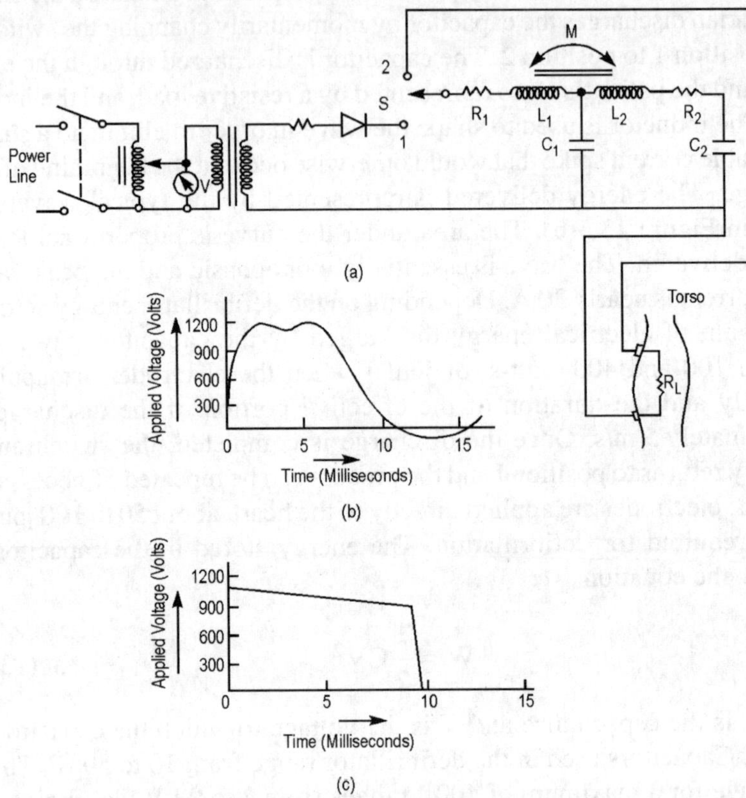

Figure 13.6 *(a) Circuit Diagram of Delay-Line Type of Capacitive Discharge Defibrillator*
(b) Waveform of the Dual Peak Defibrillator
(c) Discharge Waveform of Square-Wave Defibrillator

13.2.5 Square-Wave Defibrillator

Effective defibrillation at desirable lower voltage levels is also possible with a truncated waveform, shown in Figure 13.6(c). The capacitor is discharged through the subject by turning on a series Silicon Controlled Rectifier (SCR). When sufficient energy is delivered to the subject, a shunt SCR short-circuits the capacitor and terminates the pulse. This eliminates the long discharge tail of the waveform. The advantages of this are: (i) it requires less peak current, (ii) it requires no inductor, and (iii) it makes it possible to use physically smaller electrolytic capacitors.

13.2.6 Defibrillator Electrodes

The two defibrillator electrodes applied to the thoracic walls are called either Anterior–Anterior or Anterior-Posterior paddles. With anterior–anterior paddles, both paddles are applied to the chest. Anterior–posterior paddles are applied to both the patient's chest wall and back, so that the energy is delivered through the heart. This method of paddle application offers better control over arrhythmias that occur as a result of atrial activity. These two methods are shown in Figure 13.7(a & b). To maintain good contact, the electrodes must be firmly placed against the patient. The posterior paddle is flat and has a larger disc (with a radial handle) than the anterior paddle (axial handle). The electrodes must be sufficiently well insulated, so that the operator holding the electrodes is safe.

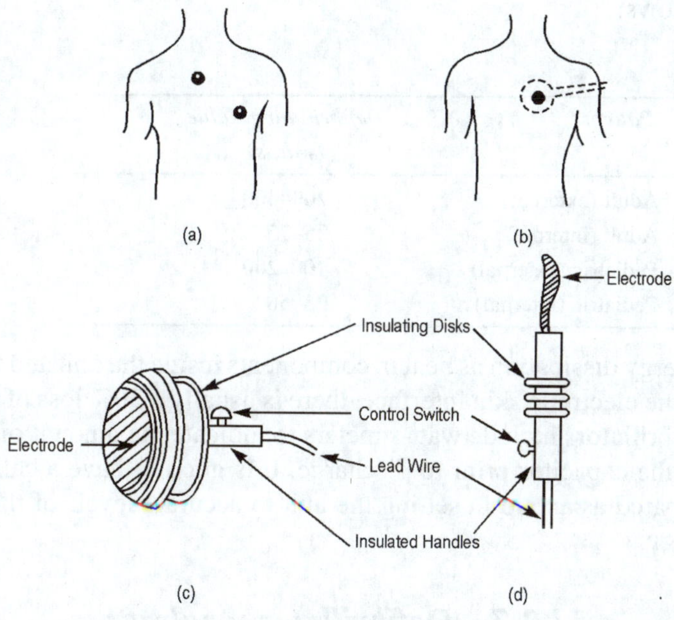

Figure 13.7 *(a) Anterior–Anterior Electrode Placement on the Chest*
(b) Anterior–Posterior Electrode Placement on Chest & Back
(c) Paddle-Type External Electrode which is Applied on Chest Wall
(d) A Spoon-shaped Internal Electrode which is Applied Directly to the Heart Muscle

Two types of electrodes for defibrillation are shown in Figure 13.7, and Figure 13.7(c) shows the type of electrode used for external defibrillation. This electrode consists of a large metal disc, approximately 100 mm in diameter, in an insulated housing. A control switch is located on the handle so that, once the electrodes are in place, the operator can push the switch to initiate the pulse. While being used, the electrodes' surface is coated with a conducting gel of the type used with an ECG recording.

Figure 13.7(d) shows an internal type of electrode which is spoon-shaped, for applying directly on the myocardium (during open-chest surgery), or it may be applied to the chest of an infant. In these applications, the energy levels required for defibrillation may range from 10 to 50 watt-s. Special pediatric paddles are available with diameters ranging from 2 to 6 cm. Internal paddles can either be gas sterilized or autoclaved.

The energy of a defibrillator is usually given in terms of watt/s, referenced across a 50 ohm resistor. Most defibrillators today have a charging capacity of 400 watt-s. Typical defibrillating values used (in watt-s) are as follows:

Table 13.1

Patient	*Defibrillating Value (watt-s)*
Adult (external)	200–400
Adult (internal)	35–75
Pediatric (external)	100–200
Pediatric (internal)	25–50

Due to energy dissipation as heat in components inside the unit and to some extent at the electrode skin interface, there is usually a 20% loss of energy. Most defibrillators include watt-s meters to indicate the amount of energy stored in the capacitor prior to discharge. It is good to have a calibration chart prepared as an aid in setting the unit to accurate levels of the delivered energy.

13.2.7 Defibrillator Analyzer

A defibrillator analyzer is basically meant to measure the energy content in the discharge pulse. It works on the principle that energy contained in a pulse of arbitrary shape and time duration is given by

$$W = \int_0^T e(t).i(t)\, dt \qquad ....(13.2)$$

where W = energy in watt-s, e(t) = voltage as a function of time, i(t) = current as a function of time, and T = time duration of the pulse.

When the voltage exists across a fixed resistance, the energy dissipated in the resistance is given by

$$W = 1/R\int_0^T [e(t)]^2\,dt \qquad \text{....(13.3)}$$

where R is the resistance of the load.

Equation (13.3) shows that a defibrillator analyzer circuit would then consist of the blocks shown in Figure 13.8. The defibrillating pulse is applied across a standard 50 ohm load and the voltage developed across it is given to a squaring circuit. The squaring circuit consists of a four quadrant multiplier followed by an operational amplifier. The output of this device is a current which is proportional to the product of the two inputs. In the squaring mode, the two inputs are connected together so that the output is a square of the input voltage. The operational amplifier acts as a current-to-voltage converter, producing an output voltage which is proportional to the output current from the multiplier.

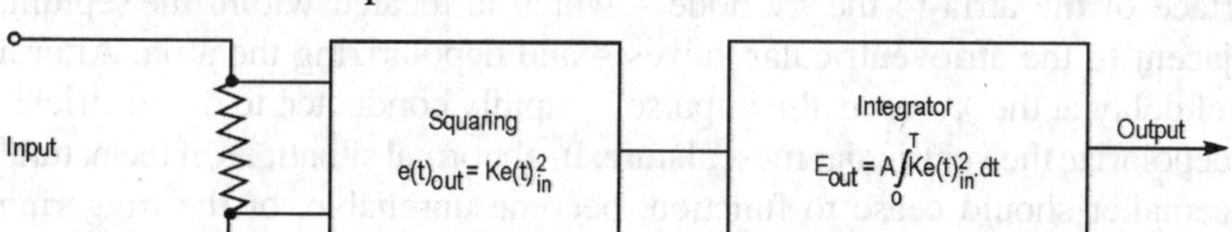

Figure 13.8 *Basic Block Diagram of a Defibrillator Energy Meter*

Besides these basic blocks, the defibrillator analyzer contains other circuits for controlling the measurement operations. For example, the pulse integrator circuit integrates the squared function pulse during the time a pulse is present. For the remainder of the display time, it acts as an analog storage element. The voltage stored in the pulse integrator is read on the energy meter calibrated in watt-s.

The instruments based on the above scheme are meant to precisely measure the pulse energy and are designed primarily for verifying or calibrating the output energy of all defibrillators. The output waveforms may be trapezoid. They measure delivered energy into a 50 ohm load, when the user places the paddles of the charged defibrillator against the input contact plates.

Besides determining the value of delivered energy from a trapezoidal defibrillator, it is often necessary to have a display of the energy waveform. This can be done by storing the waveform in digital memory and getting it out with a time expansion through a digital to analog converter, for convenient recording on a standard ECG machine.

A defibrillating pulse delivered during the T-wave period of the heart rhythm can be harmful. Hence, modern defibrillators include a provision for synchronizing the discharge pulse with the patient's ECG R-wave period. The signal is fed to an amplifier from either a patient monitor or an electrocardiograph. In some cases, the patient's ECG electrodes are connected directly to the amplifier. When properly programmed, the defibrillator will discharge only at the desired portion of the ECG waveform. The closing of the thumb switches on the paddles applied to the patient, allows the defibrillator to discharge at the next occurrence of the R wave.

13.3 Pacemakers

The rhythmic action of the heart, as we know, is initiated by regularly recurring action potentials (electrochemical impulses) originating at the natural cardiac pacemaker, located at the SA node. Each triggering pulse of the pacing impulse is propagated throughout the myocardium, spreading over to the surface of the atria to the AV node – which is located within the septum, adjacent to the atrioventricular valves – and depolarizing the atria. After a brief delay at the AV node, the impulse is rapidly conducted to the ventricles to depolarize the ventricular musculature. In abnormal situations, if the natural pacemaker should cease to function, become unreliable, or the triggering pulse does not reach the heart muscle because of blockage by the damaged tissues, the natural and normal synchronizing of heart action gets disturbed, causing the pumping action of the heart to be interfered with and leading to a seizure and possible death. When monitored, this manifests itself by a decrease in heart rate and changes in ECG waveforms. By giving external electrical stimulation impulses (Figure 13.9) to the heart muscle, it

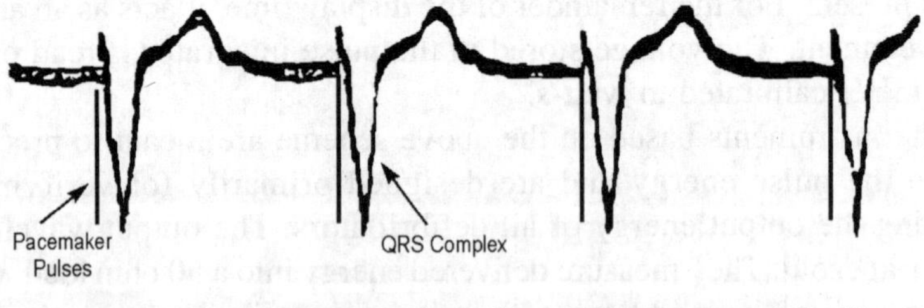

Figure 13.9 *Pacemaker Pulses Followed by QRS Complex of the Heart*

is possible to regulate the heart rate. These impulses are given by an electronic instrument called a *Pacemaker*.

Pacing is indicated for the treatment of many disorders of cardiac rhythm. Although bradyarrhythmias constitute the major indication for cardiac pacing, it is also used for the treatment of tachyarrhythmias. The disease of any part of the impulse-forming and conduction system in the heart may manifest as bradycardia. This includes pathological abnormalities in the SA node, AV node and His Purkinje system. It is now accepted that patients with any form of block arising below the AV node require pacing, because the progression of a defect with ventricular standstill is frequent and unpredictable. The presence of an acutely developing left-bundle branch block is associated with high mortality and is also regarded as an indication for pacing. A right-bundle branch block, too, may require treatment.

Cardiac pacemakers are essentially low current, low duty cycle stimulators and deliver periodic electrical impulses conducted to electrodes placed on the surface of the heart muscle (myocardium or within the heart cavity). A variety of pacemakers with their possibilities of operation are commercially available, each having some special advantages when used under particular circumstances. However, in almost all cases, the waveforms used for pacing the round-topped rectangular pulses of 1–3 m/s duration have rates adjustable from 50–150 pulses per minute.

The electrode arrangements for use with cardiac pacemakers can be both bipolar and unipolar systems. In the bipolar system, two electrodes are placed on the heart for myocardial stimulation. In the unipolar system, one electrode is placed on the heart and the other electrode is positioned elsewhere on the body. The active electrode is connected to the negative terminal of the pacemaker, so as to resemble the normal depolarization of the heart which occurs with intrinsic negative stimulation. Cardiac pacing has reached the point where there are small, reliable, long lasting units and a lot of possibilities regarding the mode of stimulation.

13.3.1 Types of Pacemakers

A device capable of generating artificial pacing impulses and delivering them to the heart is known as a pacemaker system (commonly called a pacemaker). A pacemaker basically consists of two parts: (i) an electronic unit which generates stimulating impulses of controlled rate and amplitude, known as a pulse generator, and (ii) the lead which carries the electrical pulses from the pulse generator to the heart. The lead includes the terminal which connects to the pulse generator and the insulated conductors, which interface with the electrodes and terminate within the heart.

The classification of pacemakers into different types is based on the mode of application of the stimulating pulses to the heart. *External pacemakers* are used when the heart block presents itself as an emergency and when it is expected to be present for a short time (during critical post-operative periods and in patients during cardiac surgery, involving the valves or septum). *Internal pacemakers* are used in cases requiring long term pacing because of permanent damage that prevents normal self-triggering of the heart. In the latter case, the pacemaker itself may be implanted in the body. The patient is able to move about freely and is not tied to any external apparatus. Internal pacemakers are implanted with the pulse generator placed in a surgically formed pocket below the left or right clavicle, in the left subcostal area or, in women, beneath the left or right major pectoralis muscle. Internal leads connect to electrodes that are directly in contact with the inside of the right ventricle or the surface of the myocardium (see Figure 13.10).

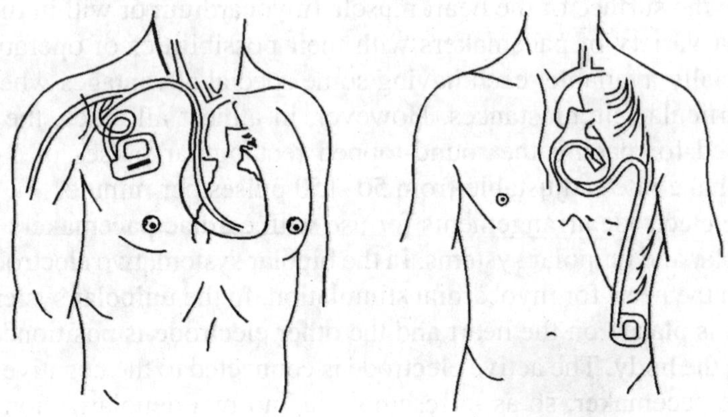

(a) (b)

Figure 13.10 *(a) Implanted Standby Pacemaker with Catheter Electrodes Inserted through the Right Cephalic Vein*
(b) Myocardial Electrodes with Pacemaker Generator Implanted in the Abdomen

The exact positioning of the electrodes depends primarily on the nature of the cardiac dysfunction, and the method (mode) of pacing that may be prescribed. Since there are no external connections for applying power, the pulse generator must be completely self-contained, with a power source capable of continuously operating the unit for a period of a few years. External pacemakers, which include all types of pulse generators located outside the body, are normally connected through the wires introduced into the right ventricle via a cardiac catheter. The pulse generator may be strapped to the leg of a patient who is confined to bed or worn at the mid-section of an ambulatory patient.

13.3.2 *Modes of Operation*

Several pacing modes are possible with both internal and external pacemakers. They can be either competitive or non-competitive. Asynchronous pacing is called competitive pacing because the fixed rate impulses may occur along with natural pacing impulses and would, therefore, be in competition with them in controlling the heartbeat. The non-competitive pacemakers are generally programmed either in the demand or synchronized mode.

13.3.2.1 *Asynchronous Pacemaker*

An asynchronous pacemaker is a free-running oscillator whose electrical pulses appear at a uniform rate, thus giving a fixed heart rate. A schematic layout of this pacemaker is shown in Figure 13.11. Each block of the layout has its own importance and must be highly reliable. The various blocks of the system are described below.

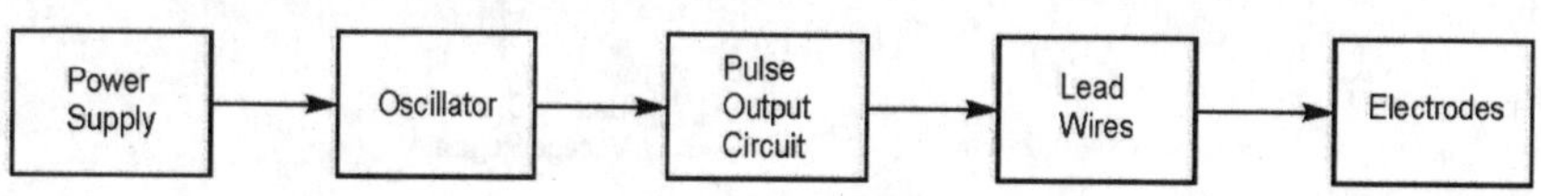

Figure 13.11 *Block Schematic Layout of an Asynchronous Cardiac Pacemaker*

Power Supply: For external pacemakers, this can be supplied through power lines which have the risk of electric shock from leakage of currents. Primary or secondary batteries giving 6 to 9 volts are therefore commonly used as a power supply source. These cells are usually discharged in two years. Lithium batteries are used nowadays which have a lifetime of ten years.

An external power source can also be used for implantable pacemakers. This is shown in Figure 13.12. The primary of the induction coil, placed close to the surface of the skin under which is a smaller internal subcutaneous coil, is packed in a harness along with its power amplifier and batterypowered RF oscillator.

The patient wears the harness before retiring at night and the charging is done due to the induction of the voltage from the external coil to the internal coil. The induced voltage is rectified, filtered and fed to a voltage regulator so that a constant voltage is provided to the implanted pacemaker even when the patient moves, resulting in a change of location of the external coil. The voltage regulator charges a battery (nickel–cadmium cells, *etc*), which powers the implanted pacemaker. One of the advantages of this system is that ordinary flashlight batteries can be used as source of power in the external circuit, which are easily replaceable.

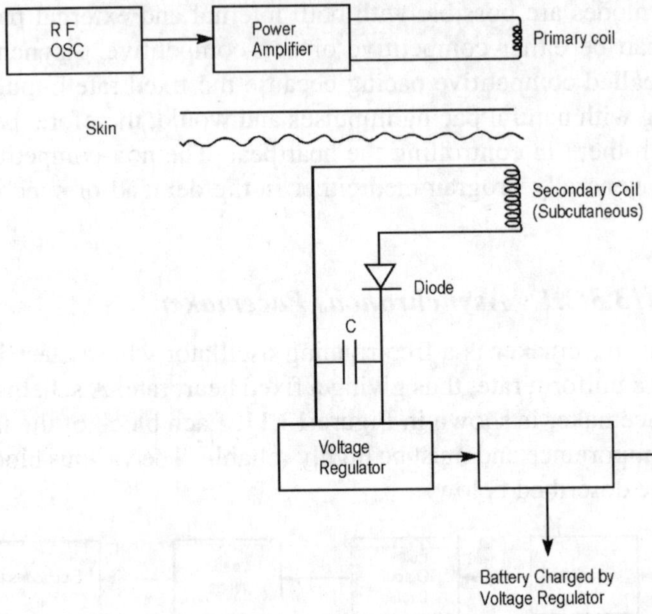

Figure 13.12 *Block Schematic of RF-Powered Cardiac Pacemaker. RF OSC, Power Amplifier and Primary Coil are Located Outside the Chest. The Rest of Circuit is Located Transcutaneously*

Oscillator: The oscillator in the asynchronous pacemaker provides stimulus pulses at a constant rate and is either a free-running blocking oscillator or a multivibrator. Figure 13.13 illustrates a blocking oscillator. When power is applied, the 3 μF capacitor is charged through the 1 MΩ resistor and reaches a voltage that forward biases the base-emitter junction of transistor Q_1, and Q_1 is thus turned ON. This results in a collector current passing through the transformer primary and an emitter current flowing through the 10 KΩ emitter resistance. The collector current in Q_1 causes current to be induced in the secondary of the transformer that discharges the capacitor and turns OFF Q_1. The process then repeats itself to produce a pulse train. The output circuit which amplifies the electrical pulse consists of a transistor operated as a switch and an appropriate coupling circuit to the electrodes. Q_2 operates as an inverter and is coupled to the electrodes through a 5 μF coupling capacitor.

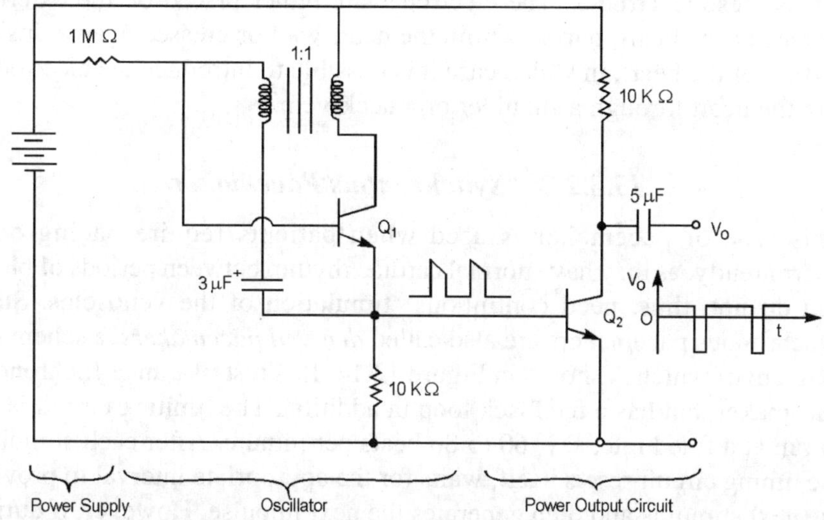

Figure 13.13 *Blocking Oscillator Frequently Used in Asynchronous Cardiac Pacemakers*

Asynchronous cardiac pacemakers produce pulses at a fixed rate between 70 to 90 beats per minute, their width ranges from 0.5 to 2 m/s and amplitude from 5 to 9 V.

Lead wires: The lead wires, besides being good electrical conductors, must also be mechanically strong to withstand stress of movements and must possess a high grade of electrical insulation. To cater to these requirements, lead wires are interwound helical coils (to minimize stress) of springwire alloy moulded in a silicon-rubber cylinder.

Electrodes: The electrodes must withstand flexing due to the pumping action of the heart and must be able to still remain in place. The material to be chosen is of great importance as it should not have electrolytic reactions with the heart tissue and also must not cause undue irritation to the myocardium. Platinum, which is chemically inert, is very frequently used for pacemaker electrodes, which are welded to special lead-wire alloys to avoid lead-wire breakage. Silver–silver-chloride is another electrode material with a low polarization potential.

Cardiac pacemakers use either unipolar or bipolar type of electrode placement. In the unipolar system, one electrode is in contact with the heart to which negative going pulses are connected from the oscillator and a large indifferent electrode is located elsewhere in the body. In the bipolar technique, two electrodes are placed on the heart and the stimulus is applied across these electrodes. The electrodes are either placed on the external surface of the heart, buried within the heart-wall or pressed against inside surface of the heart, in which case it is possible to introduce the electrodes into the heart through a shoulder or a neck-vein.

13.3.2.2 Synchronous Pacemaker

This type of pacemaker is used when patients require pacing only intermitently, as they have normal cardiac rhythm between periods of block and do not, thus, need continuous stimulation of the ventricles. Such synchronous pacemakers are also called *demand pacemakers*, a schematic diagram of which is shown in Figure 13.14. It is just like an asynchronous pacemaker, but has a feedback loop in addition. The limiting circuit is set to run at a fixed rate, say, 60 to 80 beats per minute. After each stimulus, the timing circuit resets itself, waits for the appropriate interval to provide the next stimulus and then generates the next impulse. However, if during this interval, a natural beat occurs in the ventricle, the feedback circuit detects the QRS of the ECG signal from the electrodes, amplifies it and then the amplified signal, thus obtained, is used to reset the timing circuit. It, therefore, awaits its assigned interval before producing the next stimulus.

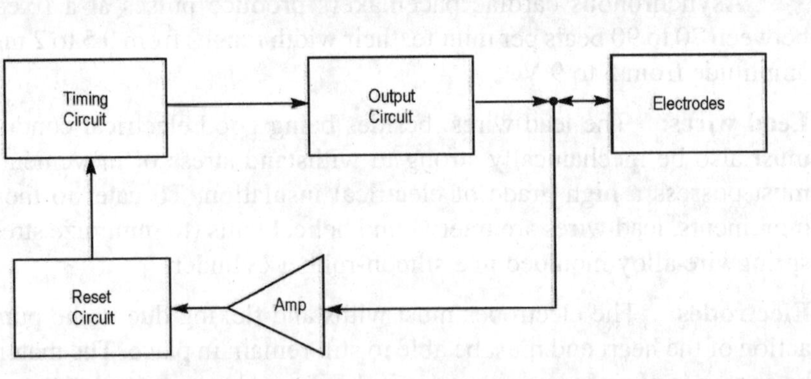

Figure 13.14 *Schematic Layout of a Demand-Type Synchronous Pacemaker*

Thus, if the heart's conduction system is operating normally with a natural rate that is greater than the rate set for the timing circuit, the pacemaker remains in standby mode. If the heart block occurs temporarily, the pacemaker takes over.

13.4 Diathermy

Applying heat to a particular area of the body increases the temperature of the tissues and the flow of blood in that area because of the dilating of blood vessels. There are several methods of raising tissue temperature, though below burning levels. One method is external, that is conductive and conductive heating, which raises skin temperature but does not penetrate very deeply. Hot compresses and the familiar infrared heat lamp are examples of this external heating. At times, it is desired that deep heating be produced without direct neuromuscular stimulation. For such cases, 'diathermy' is used. The term diathermy means 'thorough heating' or producing deep heating directly in the tissues of the body. The externally applied sources of heat like hot towels, heat lamps and electric heating pads often produce discomfort and skin burns long before adequate heat has penetrated to the deeper tissues. But with the diathermy technique, the subject's body becomes a part of the electrical circuit and the heat is produced within the body and not transferred through the skin. Another advantage of diathermy is that the treatment can be controlled precisely. Careful placement of the electrodes permits the localization of the heat to the region to be treated. The amount of heat can be closely adjusted by means of circuit parameters. Thus, using `high frequency energy', deeper lying tissues, *e.g.*, muscles, bones, internal organs, etc. can be provided with heat.

Types of Diathermy

Three basic types of diathermy are used. They are:

1. Shortwave Diathermy
2. Microwave Diathermy
3. Ultrasonic Diathermy
4. Surgical Diathermy

The first two involve electromagnetic effects and the third involves a mechanical effect.

13.4.1 Shortwave Diathermy

Here, the heating of the tissues is carried out at a high frequency of 27.12 MHz and a wavelength of 11 m. When we use currents with very high frequencies, the motor or sensory nerves are not stimulated and there is no contraction of the muscles. Thus, there is no discomfort to the patient. In this method (Figure 13.15), the output of the RF oscillator is applied to the

Figure 13.15 *Block Diagram of a Short-wave Diathermy Unit*

pair of patient electrodes. The RF energy heats the tissues and promotes the healing of injured tissues and inflammations. The power delivered by the unit is about 500 W. There are several provisions to regulate the intensity of the current passed through the patient circuit. The electrodes or pads are not directly in contact with the skin. Usually, layers of towels are interposed between the metal and the surface of the body. The pads are placed so that the portion of the body to be treated is sandwiched between them. The pads form capacitor plates and the body tissues between the pads act as a dielectric. Thus, the whole arrangement forms a capacitor. When the RF current is applied to the pads, the dielectric loss of the capacitor produces heat in the intervening tissues. This technique is called the condenser or capacitor method, Figure 13.16(a).

Further, there is also an inductive method, in which a flexible cable is coiled around the arm or knee or any other portion of the patient's body, where plate electrodes are inconvenient to use. When a RF current is passed though the cable, deep heating in the tissue results from the electrostatic field set up between its ends and the superficial tissues are heated by eddy currents set up by the magnetic field around the cable, as shown in Figure 13.16(b). This technique is known as *Inductothermy*.

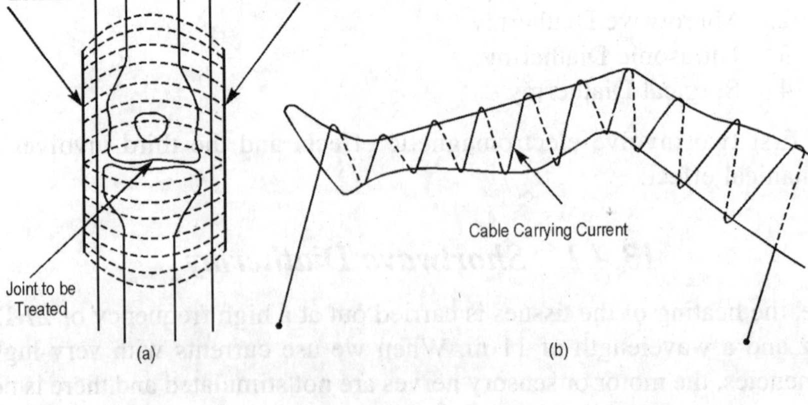

Figure 13.16 *Methods of Applying Electrodes in Short-wave Diathermy Treatment.* (*a*) *Condenser Method,* (*b*) *Inductive Method*

Instead of continuous RF waves, RF pulses of 65 μ/s (with an interval between pulses of 1600 μs) are also used. This is called Diapulse short-wave diathermy. The rate of pulsations is adjustable in steps from 80 to 600 pulses per second with the peak power of the pulse ranging from 293 to 975 W. By this method, the excess tissue fluid associated with cellular damage is reduced; the healing rate is enhanced and the depth of penetration is correctly adjusted. There is no danger of burns or hyperthermic complications.

13.4.2 Microwave Diathermy

Here, the frequency used is 2450 MHz, corresponding to a wavelength of 12.25 cm. Heating of the tissues is produced due to absorption of the microwave energy. Better therapeutic results are obtained by using microwave diathermy rather than short-wave diathermy. Here, there is no pad-shaped electrode. Instead, the microwaves are transmitted into the portion of the body to be treated directly from the director of the unit. Normally, magnetrons are used to produce microwaves.

A delay of about 3 or 4 minutes is required for the warming of the magnetron. An arrangement is made such that a lamp lights up after 4 minutes indicating that the magnetron is ready to deliver its output to the director. Proper cooling of the magnetron is provided. Further, the interference suppression filter in the circuit bypasses the high-frequency current pick-up.

Excessive dosage can cause skin burns and, in all cases, the sensation experienced by the patient is the primary guide for application. The skin should be dry as these waves are rapidly absorbed by water. The durations of irradiation generally range from 10 to 25 minutes.

Like all active devices, there are some hazards in using high-frequency energy. Patients with implanted metal objects, such as pacemakers, should not receive treatment in that area because of the possibility of overheating. Care should be exercised to restrict treatment near the eyes and to avoid having the eyes to come in the line of sight of the reflector.

13.4.3 Ultrasonic Diathermy

Ultrasonic therapy is used where short-wave treatment has failed and in cases where localization of the heat effect is desired. The ultrasonic diathermy units are very helpful in curing the diseases of the peripheral nervous system like neuritis, of the skeletal muscle system like arthritis and of skin ulcers. The heating effect is produced because of the absorption of ultrasonic energy by the tissues. The effect of ultrasonics on the tissues is a high speed vibration of micromassage. Micromassage is used in the treatment of soft tissue lesions. Ultrasonic massage is better than manual massage because of the greater depth of massage that can be obtained without any pain to the patient.

Figure 13.17 shows the simplified block diagram of an ultrasonic diathermy unit. The RF oscillator produces a high-frequency alternating current which excites the piezoelectric transducer. The transducer is made of barium titanate or lead zirconate titanate crystal with 5–6 cm^2 effective

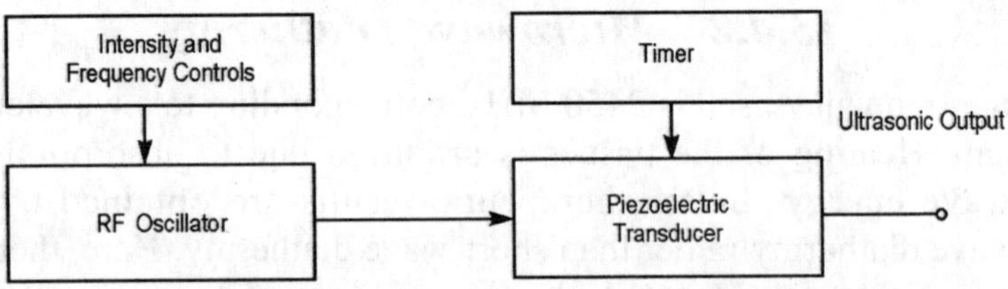

Figure 13.17 *Block Diagram of an Ultrasonic Diathermy Unit*

radiating area. The ultrasonic waves can be applied in the continuous or pulsed mode. In the case of the pulsed mode, micromassage is obtained effectively without any thermal effect. In front of the crystal, there is a metal face-plate which is made to vibrate by the oscillations of the crystal and the ultrasonic waves are emitted from this plate. The amount of ultrasonic energy absorbed by the tissues depends upon the frequency of the ultrasonic waves. Normally, a frequency ranging from 800 KHz to 1 MHz is the suitable frequency for ultrasonic therapy. The output power can be varied from 0 to 3 W/cm^2. Thus, the intensity of the ultrasonic waves is monitored in terms of electric power converted into acoustic power. The treatment timer is an electrically-operated contact which can be set from 1 to 15 minutes. Thus, it switches off the output power after the preset time. The transducer probe is in direct contact with the body through a couplant (sonic gel). In cases of large areas to be treated, the probe is moved up and down or circularly to obtain a uniform distribution of ultrasonic energy. If there is a wound or an uneven part like a joint, the treatment is carried out in a warm water bath to avoid mechanical contact with the already injured tissues. Then the probe is moved over the area at a distance or 1-2 cm from the area under treatment.

13.4.4. Surgical Diathermy

High-frequency currents, apart from their usefulness for therapeutical applications, can also be used in the operating rooms for surgical purposes involving cutting and coagulation. The frequency of currents used in surgical diathermy units is in the range of 1–3 MHz, in contrast with the much higher frequencies employed in short-wave therapeutic diathermy equipment. This is necessary to avoid intense muscle activity and the electrocution hazard which occur if lower frequencies are employed.

Surgical diathermy machines depend for their action on the heating effect of the electric current. When high frequency current flows through the sharp edge of a wire loop or band loop or the point of a needle into the tissue (Figure 13.18), there is a high concentration of current at this point. The tissue is heated to such an extent that the cells immediately under the electrode are torn apart by the boiling of thc cell fluid. The indifferent electrode establishes a large area of contact with the patient and the RF current is, therefore, dispersed so that very little heat is developed at this electrode. This type of tissue separation forms the ***basis of electrosurgical cutting***. Electrosurgical coagulation of tissue is caused by the high-frequency current flowing through the tissue and heating it locally so that it coagulates from the inside. The coagulation process is accompained by a greyish-white discolouration of the tissue at the edge of the electrode. In contrast to thermocautery, better coagulation can be achieved by high-frequency currents because this does not cause superficial burning. The concurrent use of continuous radio-frequency current for cutting and a burst wave radio-frequency for coagulation is called the *Hemostasis mode*.

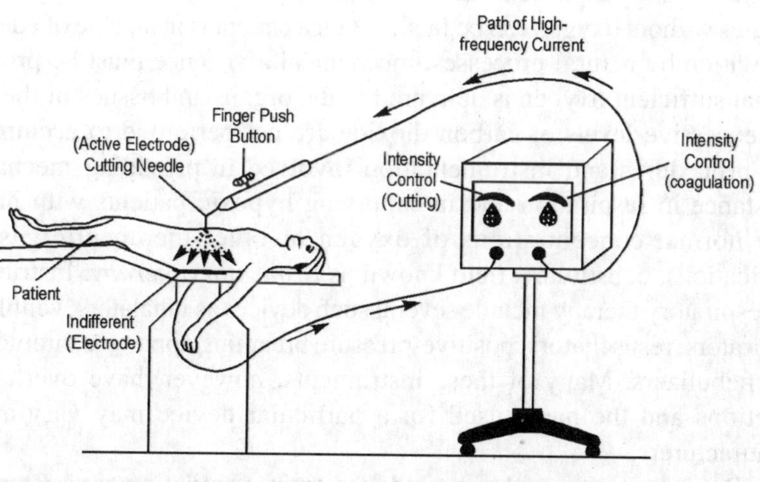

Figure 13.18 *The Principle of a Surgical Diathermy Machine*

The use of high-frequency current offers a number of interesting advantages. The separation of tissues by electric current always takes place immediately in front of the cutting edge and is not caused by it, therefore, electric current does not require any application of force. Rather, it permits elegant and effortless surgery. The electrode virtually melts through the tissue instantaneously and seals capillaries and other vessels and prevents contamination by bacteria. The simplified method of coagulation saves valuable time, since bleeding can be arrested immediately by touching the spot briefly with the coagulating electrode. A high-frequency apparatus is regarded as standard equipment in the operating theatre and it is kept ready for all surgical interventions, even if high-frequency surgery is not intended, so that its superior method of electrocoagulation can be used.

The frequency of operation of solid-state diathermy machines is 250 KHz – 1 MHz. They deliver 400 W in 500 ohms load at 2000 V in the cutting mode and around 150 W in the coagulation mode. In coagulation, the burst duration is 10–15 μ/s and repetition frequency of the burst is 15 KHz. These machines also incorporate an independent bipolar RF generator for microsurgery procedures offering a fine output power control. The output waveform is a damped sinusoid at a repetition frequency of 30 KHz. The power output across 300 ohms is about 20 W, with a peak-to-peak voltage of 1500 V open circuit.

13.5 Respirators

A vital life support which must be constantly maintained is oxygen. Five minutes without oxygen can be fatal. When a patient is incapable of adequate ventilation by natural processes, mechanical assistance must be provided so that sufficient oxygen is delivered to the organs and tissues of the body and excessive levels of carbon dioxide are not permitted to accumulate. The procedures and instrumentation involved in providing mechanical assistance in respiration and in supplying hypoxic patients with higher-than normal concentrations of oxygen or other therapeutic gases or medications, constitute a field known as *respiratory therapy*. Instruments for respiratory therapy include several such devices as inhalators, ventilators, respirators, resuscitators, positive-pressure breathing apparatus, humidifiers and nebulizers. Many of these instruments, however, have overlapping functions and the name used for a particular device may vary among manufacturers.

Respirators may either provide the entire ventilation or assist natural respiratory activity. The simplest respirator is a reciprocating pump operated by means of an electric motor. The motor operates the pump in a cyclic manner. For part of a cycle, the pump draws air from the atmosphere and pumps the air into the lungs. Later in the cycle, the air is drawn out of the lungs with a small negative pressure. The frequency of operation is determined by the speed of the motor and the coupling with the pump. No electronic device is involved in such a respirator.

The simple respirator described above is suitable only for total external ventilation. For assisting natural ventilation, the simple respirator is not suitable because there is no provision for adjustment of the rate and volume displacement in accordance with change in the natural respiratory effort. Consequently, such a respirator may at times oppose the natural effort, thereby reducing total ventilation. To properly match the external ventilatory force with the natural respiratory effort, electronic techniques are used. The respirator incorporating electronic techniques for matching is known as a 'synchronized respirator' that is, the external respiratory force is synchronized with the natural respiratory effort. Sometimes, there is a situation in which the natural respiratory rate is so low that if the external respirator synchronized with the natural rate, adequate ventilation for sustaining life would not be provided. Hence, synchronized respirators must have a provision for automatic desynchronization when the natural rate is too low. When desynchronized, the external respirator works independently of the natural system at a predetermined rate. In addition, when the natural rate approaches the value suitable for adequate ventilation, the respirator must automatically synchronize with the natural rate.

The synchronized respirator has two main functional sections. One section is the mechanical air or some other gas source. Usually, this section is in the form of a gas storage tank maintained at constant pressure by means of a pump. The second functional section of the synchronized respirator is the automatic control mechanism.

The automatic control mechanism must sense the natural respiratory activity and adjust the operation of the respirator to obtain synchronization. Several different physiological parameters have, from time to time, been used for sensing natural activity. The earliest device used the action potentials from the respiratory muscles to detect the natural activity. These potentials are present during inspiration and, in certain cases, during expiration. The potentials can be picked up with surface electrodes and amplified as is done for electromyography. The electromyographic signals cannot be directly used for control purposes. Therefore, the signals are integrated by an RC integrating circuit. The integrated signals are used to trigger an electronic pulse generator which, in turn, controls the flow of gases to the lungs. Figure 13.19 schematically shows the arrangement of the respirator sections.

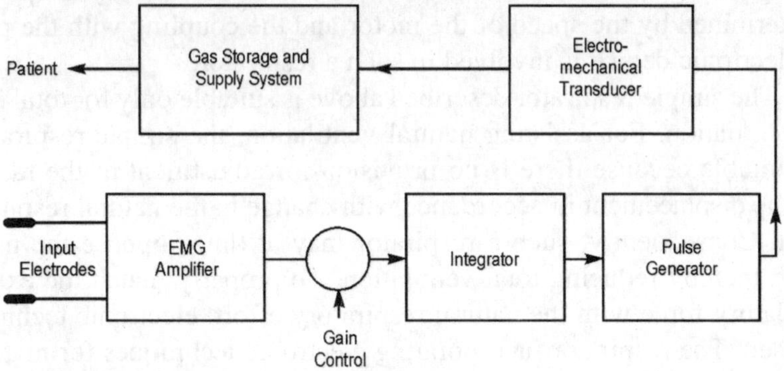

Figure 13.19 *A Synchronized Respirator System*

In order to provide desynchronization, an additional comparator must be added so that whenever the natural rate falls below a certain value, desynchronization occurs. The method, although used in respirators, is not quite satisfactory. Efforts have been made to obtain a proper physiological basis for desynchronization. The detection of carbon dioxide levels of the exhaled gases provides such an indicator. An increase of the carbon dioxide content over a certain level provides the trigger for desynchronization. The carbon dioxide content can be determined by means of an infrared CO_2 analyzer connected to the gas outflow path.

13.6 Blood Pumps

Blood pumps are required for medical research, clinical and surgical work, either to augment the natural circulation or totally take over the functions of the heart. When used to augment natural circulation, the pumps may be placed in series with the heart or in shunt with the aorta. When used in series, the blood pumped by the heart flows into the pump. The pump boosts the pressure so as to improve circulation. Such a pump is known as a *cardiac booster pump*. In the other type of circulatory assistance, known as 'arterio-arterial' pumping system, following systole, part of the blood pumped by the heart is collected in a flexible rubber or plastic chamber. Later, during diastole, the chamber is squeezed to pump out the blood into the peripheral vascular system. For totally taking over the function of the heart, the entire blood returning via the vena cava is allowed to flow into the pump. The pump then forces the blood back into the aorta. Gaseous exchange is usually performed within the pumping system.

The blood pump sometimes also known as the *artificial heart* is, therefore, essentially a mechanical device operated by means of electrical motors. For the operation of the pump to be physiologically effective, the pressure and flow must be controlled to meet the requirements of the situation. For cases of circulatory augmentation particularly, the action of the pumping system must be co-ordinated with the natural circulatory action. Electronic techniques play a significant role in control and co-ordination of the pumping process.

The electronic systems used with cardiac pumps come in a variety of forms, depending on the design of the individual manufacturer. However, all of them will have the basic elements of an automatic control system. The automatic control system will have sensing elements, comparators and actuating devices. The four most important sensing devices are for pressure, flow rate, temperature and the ECG. For pressure measurement, a transducer employing say, a strain gauge, is used. For flow rate measurement, techniques similar to the sine-wave electromagnetic flow meter can be used. If the pumps are of the positive displacement type, sometimes the flow rate can be determined by measuring the rate of movement of the piston or other movable parts of the pump. The temperature of the blood is measured with a thermistor type of thermometer. Monitoring of the ECG is important for cases where the pump has to assist cardiac activity. The R-wave of the ECG gives a good indication of the commencement of systole and, as a consequence, the R-wave can be used to synchronize the pump with the natural cardiac cycle.

The sensors are used in the control system with the objective of controlling the blood pressure, the total flow rate of blood and the temperature. The waveform of the pressure may also be controlled to correspond to a present pattern. In pumps used in combination with ventilation systems, in what is commonly known as the *heart-lung machine*, additional control for the oxygen and carbon dioxide contents of the blood may be provided. The total take-over of cardiac function by means of external pumps is resorted to only during surgical operations. For such applications, manually adjusted controls are used. For assisted circulation over relatively prolonged periods, automatic control systems are required. Since the pumping rate must match with the natural cardiac rate, the two controls possible for controlling blood pressure and flow are the displacement of the pump and the force exerted by the pump. Both the controls can be produced by means of suitable servomotors.

13.7 Myoelectric Control of Paralysed Muscles

Myoelectric control of paralysed muscles in the arm (for example) is an example of an electronic system for affecting the living system over an extended period of time. Essentially, the electronic system is prosthetic in nature.

Lack of mobility of the arm can be due to a number of causes. One possible cause is damage to the neural conducting system. For such subjects, the muscles in the arm may be intact but due to the lack of proper electrical stimuli, the patient cannot 'will' his hand to move. Not only does the situation lead to physical disability but, in addition, if allowed to persist for quite some time, will lead to atrophy of the muscles due to disuse. The paralysed arm can, however, be activated by means of electrical stimuli given from an external source. By varying the modality and locations of the stimulus given, different types of movement can be induced in the arm. From the practical viewpoint, such alteration of modality and location of the stimuli is not convenient. Alternatively, the characteristics of the stimulus given to the arm can be varied over a wide range to produce the desired motion. The most convenient way to vary the characteristics is to obtain command signals from some other muscles. For subjects who have only one arm paralysed, the command signals may be obtained from the intact arm. Since the EMG signals from an intact arm may have a wide range of variations, depending on the movement 'willed' to that arm, the command signals may have adequate variations to give a number of desired movements to the paralysed arm. However, one-to-one correspondence of movement of the two arms is not yet possible.

The functional electronic system arrangement used is schematically shown in Figure 13.20. For a patient, the EMG voltages are picked up by means of surface electrodes placed on the intact arm. The EMG signals are

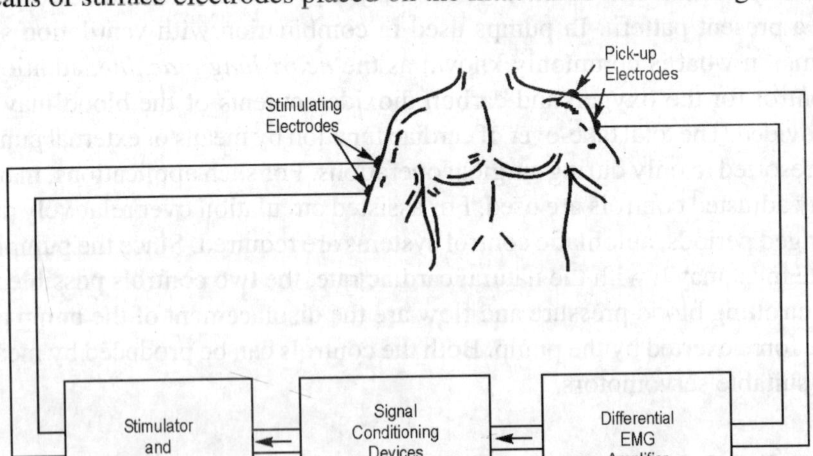

Figure 13.20 *Schematic Representation of a Myoelectric Control System*

amplified using compact and powerful amplifiers with characteristics similar to the electromyograph amplifiers described in Chapter 6. The EMG signals cannot be used directly for stimulation and, therefore, signal conditioning techniques have to be adopted. One simple form of signal conditioning is to rectify the amplified EMG signal using a diode rectifier. The rectified signal is then integrated using a short-time, constant RC integrator. As a result, a wave corresponding closely to the envelope of the signal is obtained. The envelope is a DC signal with relatively slow variations. Hence, effective stimulation cannot be obtained with this voltage. Consequently, an additional stimulator is used. The integrated EMG signal is used to modulate the stimulator output. For simplicity, only amplitude modulation may be used. For greater effectiveness, a combination of amplitude, pulse width and repetition rate modulation is used. The modulated output of the stimulator is fed to the electrodes fixed to the paralysed arm. Thus, a movement of the intact arm causes signals of varying characteristics to be fed to the modulator. The stimulator output changes accordingly and the muscles of the paralysed arm receive stimulation of various characteristics. The movement of the arm in various modes can, therefore, be obtained.

One significant problem with the myoelectric control set-up is the feedback problem. The stimulation voltages are of the order of 30 V, whereas the EMG signals picked up are in the range of microvolts or a few millivolts. Since the tissue inside the body is conductive, the stimulating voltage can be partially transmited to the pick-up electrodes. Even if there is considerable loss in the transmission, since the ratio of voltages is high, significant feedback may result. Under certain circumstances, violent oscillations of the limbs may result. The current passing from the stimulating to pick-up electrode may even be injurious to the heart. One method of avoiding this hazard is to have an isolated output for the stimulators, so that there are no continuous electrical paths from the stimulating to the pick-up electrodes.

Chapter 14

Biotelemetry

In many situations, it becomes necessary to monitor physiological events from a "distance". To quote a few applications:

1. Radio frequency transmissions for monitoring the health of astronauts in space.
2. Patient monitoring, where freedom of movement is desired, such as in obtaining an exercise ECG. In this instance, the requirement of trailing wires is both cumbersome and dangerous.
3. Patient monitoring in an ambulance and in other locations away from the hospital.
4. Collection of medical data from a home or office.
5. Research on unrestrained and unanesthetized animals in their natural habitat.
6. Use of telephone links for the transmission of ECGs or other medical data.
7. Special internal techniques, such as measuring pH or pressure in the gastrointestinal tract.
8. Isolation of an electrically susceptible patient from power-line operated ECG equipment, to protect him from accidental shock.

These applications have indicated the need for systems that can adapt the methods of measuring physiological variables to a method of transmission of resulting data. This is the branch of biomedical instrumentation known as 'Biomedical Telemetry' or 'Biotelemetry'.

Biotelemetry, literally, is the measurement of biological parameters over a "distance". The means of transmitting the data from the point of generation to the point of reception can take many forms. Perhaps the simplest application of the principle of biotelemetry is the stethoscope, whereby heartbeat sounds are amplified acoustically and transmitted through a hollow-tube system to be picked up by the ear of the physician for interpretation. The purpose of this chapter is merely to outline the elements of the subject and to present an example of its application. For comprehensive treatment, the reader is referred to the Bibliography.

14.1 Physiological Parameters Adaptable to Biotelemetry

Just as with hardware systems, measurements can be applied to two categories:

1. Bioelectrical parameters, such as ECG, EEG and EMG.
2. Physiological variables that require transducers, such as blood pressure, gastrointestinal pressure, blood flow and temperatures.

With the first category, a signal is obtained directly in electrical form. The second category requires some type of excitation, because the physiological parameters are eventually measured as a variation of resistance, inductance or capacitance. The differential signals obtained from these variations can be calibrated to represent pressure flow, temperature and so on, since some physical relationships do exist.

One example is ECG telemetry, the transmission of ECGs from an ambulance or site of emergency to a hospital. A cardiologist at the hospital can immediately interpret the ECG, instruct the trained rescue team in their emergency resuscitation procedures and arrange for any special treatment that may be necessary upon the patient's arrival at the hospital. In this application, the telemetry to the hospital is supplemented by two-way voice communication.

Telemetry of EEG signals has also been used in studies of mentally disturbed children. The child wears a specially designed "spaceman's helmet" with built-in electrodes, so that the EEG can be monitored without traumatic difficulties during play. In one clinic, the children are left to play with other children in a normal nursery school environment. They are monitored continuously while data are recorded.

The third type of bioelectric signal that can be telemetered is the EMG. This parameter is particularly useful for studies of muscle damage, partial paralysis problems and also in human performance studies.

One important application of telemetry is in the field of blood pressure and heart-rate research in unanesthetized animals. The transducers are surgically implanted with leads brought out through the animal's skin. A male plug is attached post-operatively and later connected to the female socket contained in the transmitter unit.

Blood flow has also been studied extensively by telemetry. Both the Doppler-type and electromagnetic-type transducers can be employed.

The use of thermistors to measure temperature is also easily adaptable to telemetry. In addition to the continuous monitoring of skin temperature or systemic body temperature, the thermistor system has been found to be of use in obstetrics and gynecology. Long term studies of natural birth control by monitoring vaginal temperature, have incorporated telemetry units.

One more application is the use of the "radio pill" to monitor stomach pressure or pH. In this application, a pill that contains a sensor plus a miniature transmitter is swallowed and the data are picked up by a remote receiver and recorded.

It is interesting to note that biotelemetry studies have been performed on dogs, cats, rabbits, monkeys, deer, turtles, snakes, horses, seals and giraffes, as well as on humans.

14.2 Elements of Biotelemetry System

The essential blocks of a biotelemetry system are shown in Figure 14.1. The transducer converts the biological variable into an electrical signal. The signal conditioner amplifies and modifies this signal for effective transmission. The transmission link connects the signal input blocks to the read-out device by wire or wireless means.

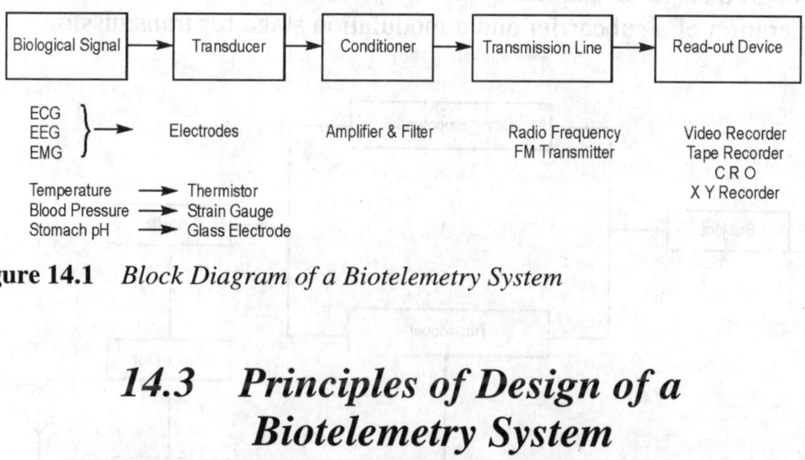

Figure 14.1 *Block Diagram of a Biotelemetry System*

14.3 Principles of Design of a Biotelemetry System

a) The telemetry system should be selected to transmit the bioelectric signals with the maximum fidelity and simplicity.
b) There should not be any constraint on the living system due to the installation of these systems and no reaction or interference from living system is desirable.
c) The size and weight of the telemetry system should be compact. For long term use units, the size and weight recommended is of the order of 1% of that of the living system. For a short duration, it is about 5%.
d) Stability and reliability are essential.

e) The power consumption should be very small to extend the source lifetime in the case of implanted units.
f) For wire transmission, the shielding of the cable is a must to reduce noise levels. At the transmitter side, the amplifiers should be differential amplifiers to reject common mode interference.
g) Miniaturization of the radio telemetering system helps to reduce noise.

14.4 Biotelemetry—Single Channel Radio Telemetry Systems

There are many types of commercial biotelemetry systems available today. To illustrate the basic principles involved in telemetry, a simple system is described. The stages of a typical biotelemetry system can be broken down into functional blocks, as shown in Figure 14.2 for the transmitter and in Figure 14.3 for the receiver. Physiological signals are obtained from the subject by means of appropriate transducers. The signal is then passed through a stage of amplification and processing circuits, that include the generation of a subcarrier and a modulation stage for transmission.

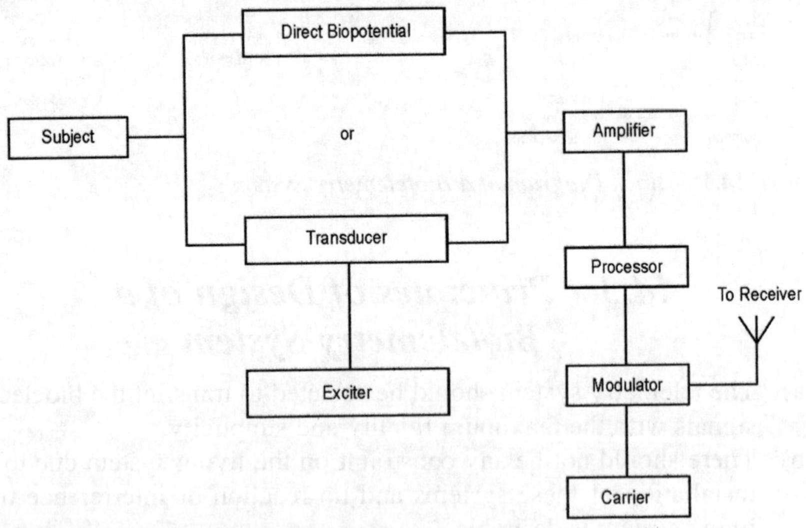

Figure 14.2 *Block Diagram of a Biotelemetry Transmitter*

The receiver consists of a tuner (Figure 14.3) to select the transmitting frequency, a demodulator to separate the signal from the carrier wave and a means of displaying or recording the signal. The signal can also be

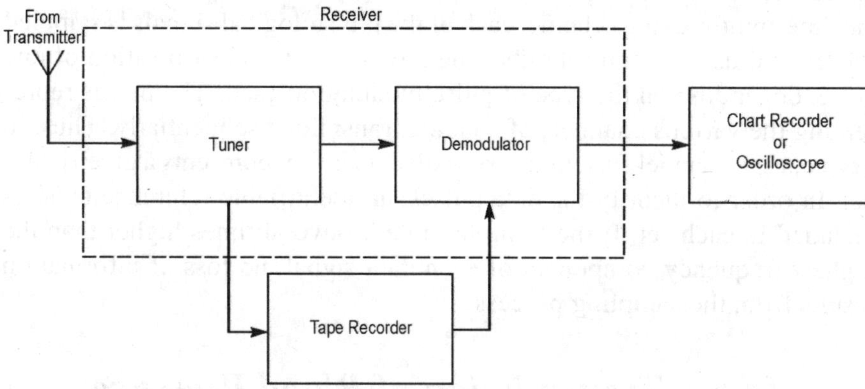

Figure 14.3 *Receiver–Storage Display Units*

stored in the modulated state by the use of a tape recorder, as shown in the block diagram.

Either amplitude modulation or frequency modulation can be used in the system mentioned above. Because of reduced interference, FM transmission is often used for telemetry. In biotelemetry systems, the physiological signal is sometimes used to modulate a low-frequency carrier, called a subcarrier, often in the audio frequency range. The RF carrier or the transmitter is then modulated by the subcarrier. If several physiological signals are to be transmitted simultaneously, each signal is placed on a subcarrier of a different frequency and all the subcarriers are combined to simultaneously modulate the RF carrier. This process of transmitting many channels of data on a single RF carrier is called 'frequency multiplexing', and is much more efficient and less expensive than employing a separate transmitter for each channel. At the receiver, a multiplexed RF carrier is first demodulated to recover each of the separate subcarriers, which must then be demodulated to retrieve the original physiological signals. Either FM or AM can be used for impressing data on the subcarriers and this may or may not be the same modulation method that is used to place the subcarriers on the RF carrier. In describing this type of system, a designation is given, in which the method of modulating the subcarriers is followed by the method of modulating the RF carrier. For example, a system in which the subcarriers are frequency-modulated and the RF carrier is amplitude-modulated is designated as FM/AM. An FM/FM designation means that both the subcarriers and the RF carrier are frequency-modulated. Both FM/AM and FM/FM systems have been used in biotelemetry, the latter more extensively.

As in AM and FM systems, the multiplexing of several channels of physiological data can be accomplished in a pulse modulation system.

However, instead of frequency multiplexing, time multiplexing is used. In the time-multiplexing scheme, each of the physiological signals is sampled briefly and used to control either the amplitude, width or position of one pulse, depending on the type of pulse modulation used. The pulses representing the various channels of data are transmitted sequentially. Thus, in a typical six-channel system, every sixth data pulse represents a given channel. In order to identify the data pulses, an identifiable reference pulse is included in each set. If the sampling rate is several times higher than the highest frequency component of each data signal, no loss of information results from the sampling process.

14.5 Transmission of Blood Pressure by Telemetry

A system for monitoring blood pressure is used to illustrate the FM/FM method of transmission. The transducer used in this case is the flush diaphragm-type of strain-gauge transducer. Electrically, it can be represented by the bridge circuit of Figure 14.4. Resistors R_1 and R_3 decrease, whereas R_2 and R_4 increase in value as blood pressure increases. Resistor R_b is simply for balancing or zeroing. The transducer is connected in the transmitter circuit as shown in Figure 14.5.

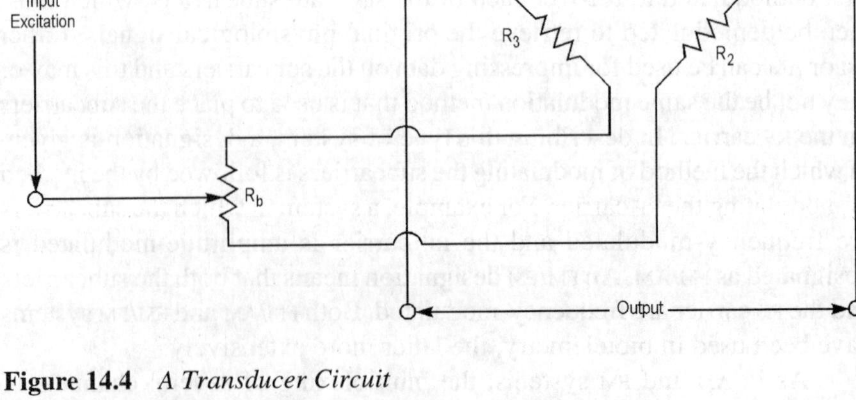

Figure 14.4 *A Transducer Circuit*

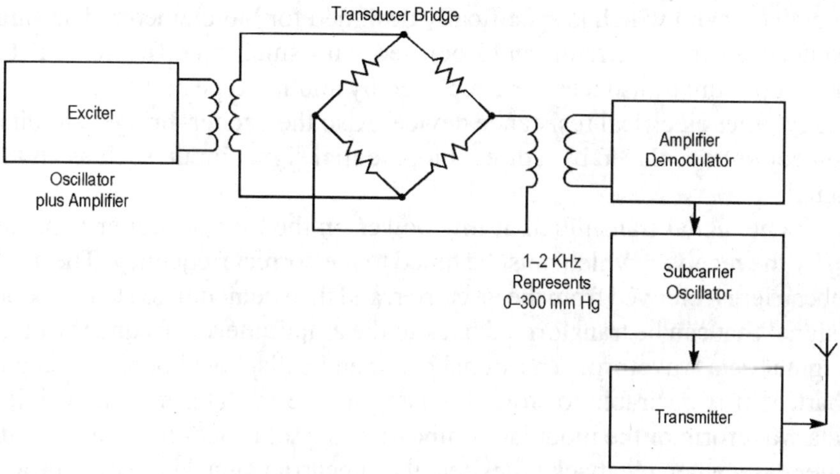

Figure 14.5 *One Type of Exciter–Transmitter Unit for Blood Pressure Telemetry*

Either DC or AC can be used as excitation for strain-gauge bridges. When DC is used, the amplifier following the bridge must be a DC amplifier, with its associated problems of stability and drift. When AC is used, the bridge acts as a modulator and filters are required in order to recover the signal.

The exciter unit consisting of a Colpitts oscillator plus an RC coupled amplifier stage, excites the bridge with a constant voltage at a frequency of approximately 5 KHz, and is coupled to the bridge inductively. The bridge is initially balanced both resistively and capacitively so that any changes in the resistance of the arms of the bridge, due to changes in pressure on the transducer, will result in changes of the output voltage. The output voltage of the bridge is inductively coupled to a class A amplifier, the output of which is RC coupled to a class C (power) amplifier. The resulting output wave is rectified to obtain a signal representative of the pressure variation. The rectified wave is put through a resistance–capacitance filter, and the resulting voltage controls the frequency of a UJT oscillator. This is the FM subcarrier oscillator that is used to modulate the main carrier.

The system can be arranged so that there is a fairly linear relationship between the subcarrier oscillator frequency and the physiological parameter to be measured. For example, in the system for blood pressure illustrated in Figure 14.5, a frequency range of 1 to 2 KHz represents the range of 0 to 300 mm Hg (0 to 40 KPa) pressure. The transducer action can be traced very easily. The subcarrier is used to frequency-modulate the main transmitter carrier. The carrier is transmitted at lower power on a frequency band which is specifically designed for biotelemetry. The same exciter-transmitter circuit could be used with small modifications, if the blood pressure transducer were replaced by another type or by a thermistor or any other electrical resistance device. Also the exciter–bridge combination could be replaced by a direct biopotential signal input, such as an ECG signal.

The signal transmitted at low power on the FM transmitter is picked up by the receiver, which must be tuned to the correct frequency. The audio subcarrier is removed from the RF carrier, and then demodulated to reproduce a signal that can be transformed back to the amplitude and frequency of the original data waveform. The signal can then be displayed or recorded on a chart. If it is desirable to store the data on tape for later use, the original data waveform or the modulated subcarrier signal is put on the tape. In the latter case, when playback is desired, the subcarrier signal is passed through the FM subcarrier demodulator.

14.6 Transmission of Electrocardiogram by Telemetry

Figure 14.6 shows the block diagram of a single channel telemetry system suitable for the transmission of ECG. There are two main parts: the telemetry transmitter which consists of an ECG amplifier, subcarrier oscillator and a UHF transmitter (along with dry cell batteries), and a telemetry

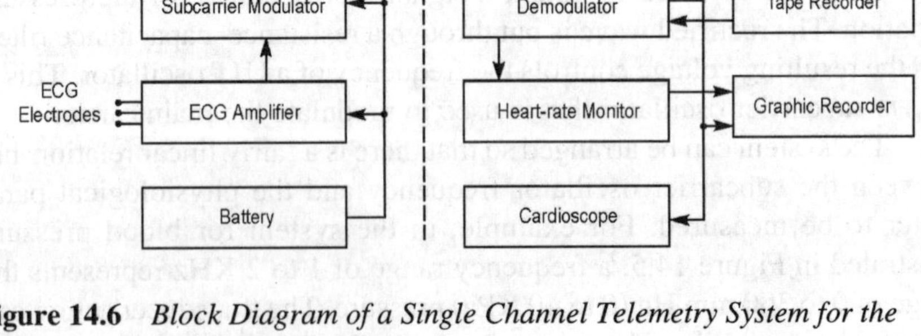

Figure 14.6 *Block Diagram of a Single Channel Telemetry System for the Transmission of ECG*

receiver, consisting of a high frequency unit and a demodulator, to which an electrocardiograph can be connected to record a cardioscope to display and a magnetic tape recorder to store the ECG. A heart-rate meter with an alarm facility can be provided to continuously monitor the beat-to-beat heart rate of the subject.

It is suggested that for distortion-free transmission of ECG, the following requirements must be met:

- The testee should be able to carry on with his normal activities, whilst carrying the instruments, without the slightest discomfort. He should be able to forget their presence after some minutes of application.
- Motion artefacts and muscle potential interference should be kept to a minimum.
- The battery life should be long enough so that a complete experimental procedure may be carried out.

While monitoring paced patients for ECG through telemetry, it is necessary to reduce pacemaker pulses. The amplitude of the pacemaker pulses can be as large as 80 mV compared to 1–2 mV, typical of the ECG. The ECG amplifier in the transmitter is slow and limited, so that the relatively narrow pacemaker pulses are substantially reduced in amplitude.

14.7 Telephone Links

Although it cannot be considered to be radio telemetry, the use of the telephone system to transmit biological data is becoming quite common. One application involves the transmission of ECGs from heart patients and (particularly) pacemaker recipients. In this case, the patient has a transmitter unit that can be coupled to an ordinary telephone. The transmitted signal is received by telephone in the doctor's office or in the hospital. Tests can be scheduled at regular intervals for diagnosing the status and potential problems indicated by the ECGs.

Figure 14.7 shows a typical telephone telemetry system for transmitting and receiving medical signals. For frequency modulation, a reactance modulator has been used with a centre frequency of 1500 Hz. This frequency is modulated ±200 Hz for a 1 V peak-to-peak signal. This deviation is linear within a 1% range. The demodulator consists of an audio amplifier, a carrier rejection filter, and a low pass integrator output circuit to recover the input signal. Both at the transmitting as well as receiving ends, coupling or isolation transformers are used to match the standard telephone line impedance.

Telephone lines can have noise interference. This noise will produce significant artefacts in the AM and FM modulation schemes described above.

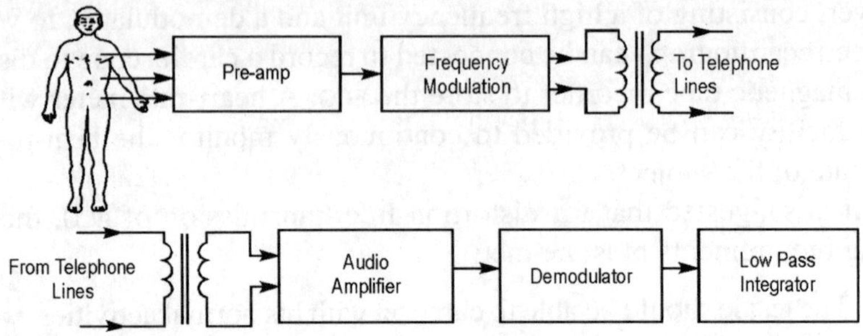

Figure 14.7 *Arrangement for the Transmission of Analog Signals over Telephone Lines*

Therefore, the current practice is to use a computational facility for on-line error correction. The scheme is given in Figure 14.8. A set of transducers and electrodes access biological data such as for temperature, breathing, heart sounds and ECG. The data is generally in analog form and in the data acquisition system (DAQ), by means of a high resolution (12-bit) analog-to-digital converter, the data is converted to digital form. A microprocessor-based system can perform some artefact rejection tasks such as nullifying drift and blocking artefactual signals which are well beyond the known range for the biomedical signal being monitored. Also, the microprocessor

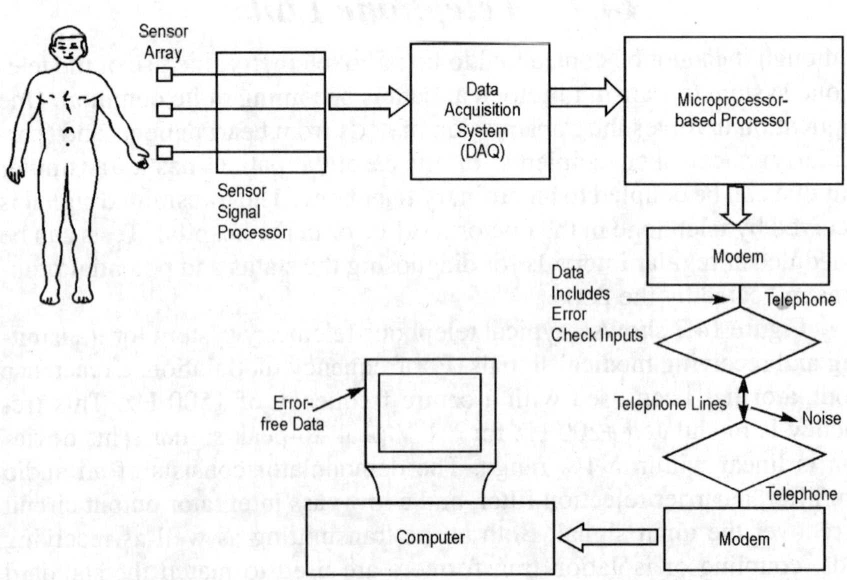

Figure 14.8 *Scheme of Medical Data Communication through Telephone Lines with Error Correction*

can time multiplex data from the different transducers, so that the information from each of the transducers is transmitted through one communication channel in the time-shared mode.

The data, now in digital form, becomes the input for a modem (acronym for Modulator–Demodulator). Modems interface the data into telephone lines according to standard protocols as, for example, TCP/IP protocol. Data transmission at 56 kbps and higher is possible. Serial, binary, asynchronous data formats are commonly used with trellis-coded modulation.

Also, an error correction facility is provided. For example, a parity check is obtained by adding to every character (formed by 7 bits) one more bit (0 or 1). At the receiving end, the parity bits are again counted and, if there is a mismatch from the input parity count, an error is indicated and that particular small packet of data is dropped. Additional error checks are commonly incorporated to ensure that the data received is error-free.

From the transmitting telephone, the data goes to the receiving telephone. A modem at the doctor's end, which is the receiving end, converts this data into a form which can be put through the serial part of the computer. The data is decoded by means of an algorithm and displayed in the computer monitor in numerical form as, for example, in the case of temperature and graphical form for signals like the ECG.

Chapter 15

Patient Monitoring and Intensive Care System

The need for intensive care and patient monitoring has been recognized for centuries. The 24-hour nurse for the critically ill patient has, over the years, become a familiar part of the hospital scene. But only in the last few years has equipment been designed and manufactured that is reliable and sufficiently accurate to be used extensively for patient monitoring. Nurses are still there, but roles have changed somewhat, for they now have powerful tools at their disposal for acquiring and assimilating information about the patients under their care. They are therefore able to render better service to a larger number of patients and are better able to react promptly and properly to a larger number of patients and to an emergency situation.

In modern hospitals, the patient monitoring system is becoming increasingly popular. The system involves continuous measurement of physiological parameters like respiratory rate, body temperature, blood pressure and heart rate. For special situations, additional monitoring of the EEG, ECG, EMG, phonocardiogram and galvanic skin response may be undertaken. The use of continuous monitoring can significantly improve the attention given by hospital staff to a patient in the most critical of conditions, and yet without neglecting other patients. In the intensive care system, patient monitoring is combined with devices for resuscitation and other specific therapeutic equipment.

Most patient monitoring equipment and intensive care systems are of modular form. By changing plug-in units, a variety of functions can be obtained. Two basic requirements must be met. First of all, for any desired parameter, continuous monitoring should be possible and, secondly, a system of alarm must be provided so that if a particular parameter changes to a value indicating a critical stage, the doctor or nurse must be warned of it. In large systems, continuous monitoring is not possible because too many display devices would be required. In such a case, provision must therefore be made for automatic scanning of data from individual beds. The transducers and pick-ups used must be rugged and extremely reliable. To obtain reliability, transducers and pick-ups are usually made as simple as possible, even at the sacrifice of sensitivity and speed of response. For example, for respiration rate, instead of a spirometer type of equipment, a thermistor placed near the nasal opening is used. When the patient breathes out, the variation of resistance can give an indication of the respiratory rate and, to some extent, the depth of respiration. For body pressure measurement, the strain-gauge transducer is usually not suitable because prolonged insertion of a needle or catheter might be painful and can also have adverse psychological effects. Instead, for this purpose, the photoelectric finger pulse monitor or the photoelectric plethysmograph is used.

The photocell transducer can also be used to determine the heart-rate. Since most monitoring systems have a provision for ECG, it is more convenient to automatically determine the heart-rate from the ECG R-R wave interval (*vide* Chapter 8).

15.1 System Arrangement

One possible system of arrangement is to have bedside monitors consisting of individual units which may be wall-or shelf-mounted, set on mobile stands or hung from the ceiling. The monitoring equipment, for example, in the intensive coronary care area is installed beside the bed of each patient to measure and display the ECG, heart-rate, and other parameters being monitored from that patient. In addition, information from several bedside stations is usually displayed on a central console at the nurses' station. One type of intensive coronary care unit layout that is quite popular is a U-shaped design in which six or eight cubicles or rooms with glass windows surround the nurses' central monitoring station. Although the optimum number of stations per central console has not been established, a group of six or eight seems most efficient. For larger hospitals, the monitoring of 16 or 24 beds can be accomplished by two or three central stations. The exact number depends on the individual hospital, its procedures, and the physical layout of the patient-care area. In certain areas in which recruitment of trained nurses is difficult, this factor could also be considered in selecting the best design.

Although patient monitoring systems vary greatly in size and configuration, certain basic elements are common to nearly all of them. A cardiac care unit, for example, generally includes the following components:

1. Skin electrodes to pick up the ECG potentials.
2. Powerful ECG amplification equipment.
3. A CRT display permitting direct observation of the ECG waveform, incorporated on the bedside monitor (with a screen of 5 to 12.5 cm

diameter), while the central nurses' station generally has a larger screen on which ECGs of several patients are displayed simultaneously.

4. A rate meter used to indicate the average number of heartbeats per minute and to provide a continuous indication of the heart-rate.
5. An alarm system, actuated by the rate meter, to alert the nurse or another observer by an audible or visible signal whenever the heart-rate falls below or exceeds some adjustable preset range, (*e.g.*, 40 to 150 beats per minute).

In advanced versions of the intensive care system, as in Figure 15.1, the central station is also provided with data recording tape recorders. The data is recorded in digital form so that later on they may be directly analyzed with the help of computers. Each type of data from the bedside monitors is fed to an electronic threshold comparator. The comparator senses the level of the signal and whenever the level falls too low or becomes too high, an alarm is given. The alarm may be in the form of blinking lights or bells or a combination of both. The system should be so arranged that from the alarm signal it should be possible to indentify the bed with the critical subject and the particular physiological parameter that has become abnormal.

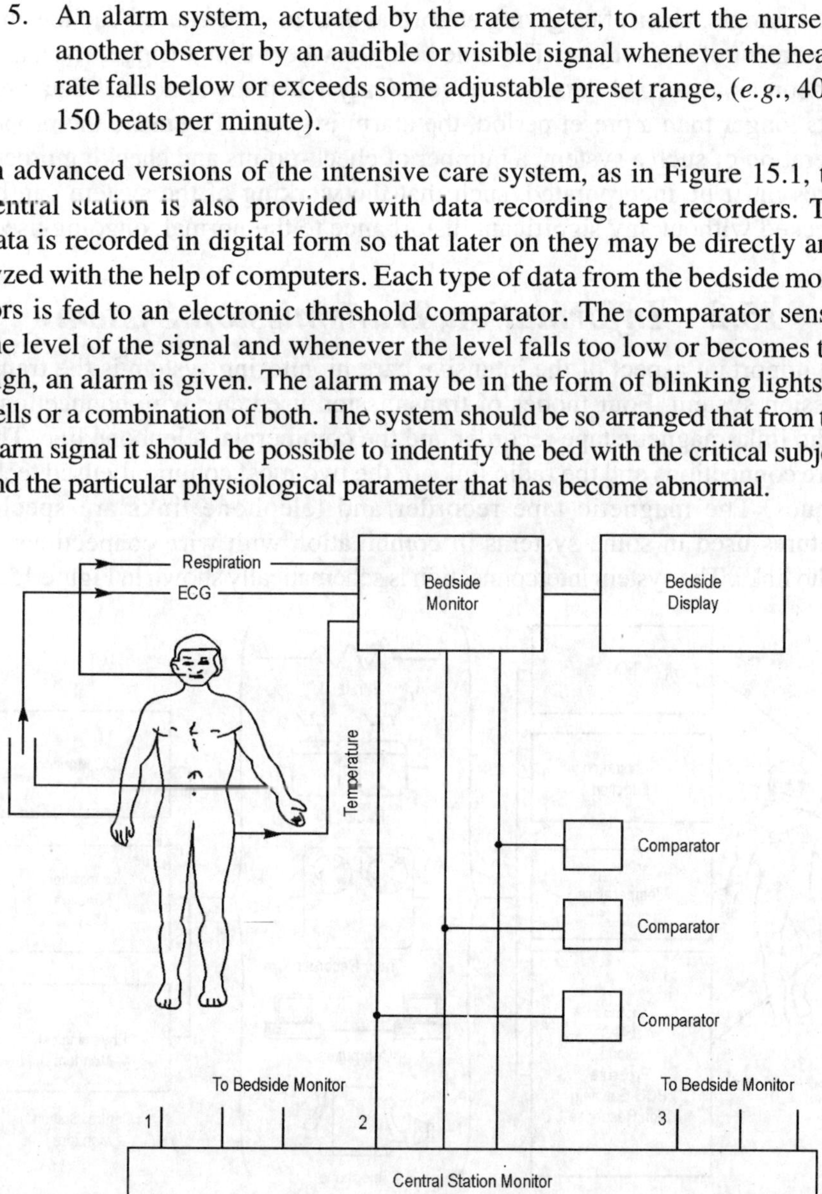

Figure 15.1 *System Arrangement of an Intensive Care Unit (ICU)*

Experience with the use of patient monitoring systems has shown that false alarms are frequent, particularly in the initial stages of operation after installation. In addition, faults frequently occur in parts of the system and they do not give the warning signal when they are required to do so. Therefore, many patient monitoring systems have a built-in time delay. For example, an artefact giving abnormal data may be caused due to some movement of the subject. The time delay device will not trigger the alarm as soon as abnormal data is received. Only when the abnormal data persists longer than a preset period, the alarm is given. To ensure the proper operation of such a system, a number of check-points and checking procedures must be incorporated, such that the working of the system can be checked without any significant disturbance to the normal, ongoing use.

15.2 Information Transmission System

An important aspect of the intensive care monitoring system is the transmission system. Four modes of transmission used are: wire connections, radio link, magnetic tape recorder and the commercial telephone line. The wire connections and the radio link are the two most commonly used techniques. The magnetic tape recorder and telephone links are special features used in some systems in combination with wire connections or radio links. The system interconnection is schematically shown in Figure 15.2.

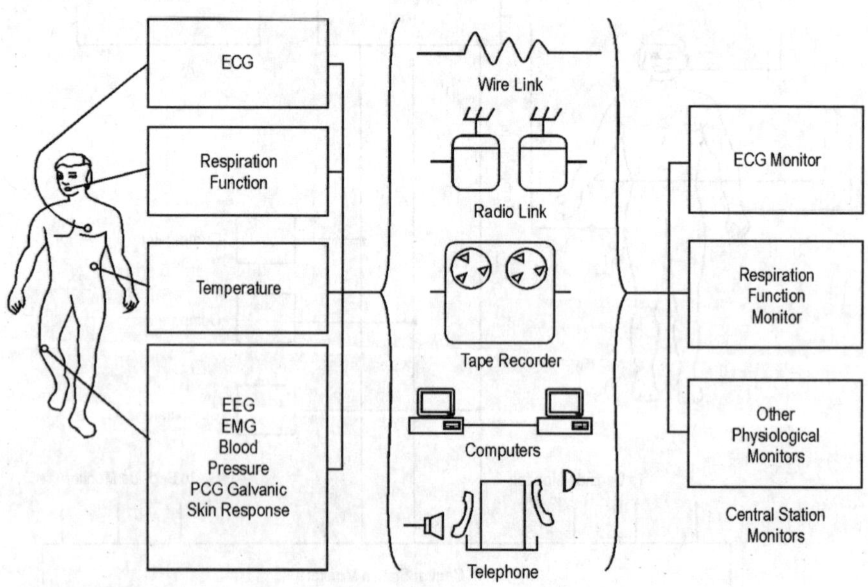

Figure 15.2 *Schematic Representation of a Data Transmission System in an ICU*

The wire connections have the advantages of simplicity and low cost. However, the installation procedures are cumbersome and require considerable structural alterations, especially if a system has to be installed in an existing hospital room. Radio links are useful for situations where the intensive care system has to be installed in an existing hospital room or where the arrangments of the beds and the central monitor may have to be altered from time to time. When miniature transmitters, one for each physiological parameter monitored, are used the patient gets a feeling of freedom. The transmitter can be strapped onto the patient, and since no connecting leads are used, the patient's movements are not restricted. Individual transmitters for each physiological parameter are not always economical or feasible. In such cases, all physiological parameters are fed to a bedside transmitter, which transmits all the parameters in a time-shared fashion. The transmitters used usually have a range of about 100 m. The frequencies must be according to permission obtained from the concerned government authorities. The radio link system, however, suffers from the problem of electrical interference and the system may have poor performance under certain atmospheric conditions.

The magnetic tape recorder as a system of interconnecting devices is not a convenient system by itself, because it involves tape transport. However, the magnetic tape recorder is frequently used with other interconnection techniques to provide permanent records and playback facilities.

The telephone link is also not suitable as an interconnection system by itself. But it is gaining in popularity for the interconnection of an individual intensive care system, with a central information analysis centre. For example, a number of ICUs spread over a certain zone of the country may be connected by telephone links to a central information and analysis section. For critical and complicated cases, the data received can be sent to the central station for analysis and for advice. The analysis and advice may be handled by a computer.

With the current ready availability of computerized local area networks (LAN) and, on a larger scale, fast Integrated Services Digital Network (ISDN) communication systems, a large part of the data communication is now computer-based. Within the hospital as well as between quite geographically separated hospitals in real time, dynamic data can be transferred. Also, the data-handling capacity has so increased that images can also be transferred.

15.3 Resuscitation Unit

One or more resuscitation units is a must for every intensive care system. In modern hospitals, the resuscitation units are in the form of a mobile trolley on which all the essential items are mounted. The units mainly consist of cardiopulmonary support devices, including respirators and cardiac massage devices.

The main electronic equipment mounted on resuscitation trolleys are:

1. A single or dual trace monitoring oscilloscope for ECG and pulse monitoring.
2. An instantaneous heart-rate meter operating both from the ECG and the arterial pulse.
3. A graphic recorder for direct paper write-out of the oscilloscope trace.
4. A DC defibrillator for internal and/or external use.
5. A synchronizer for synchronization of the external pacemaker with the natural ECG.
6. A pacemaker for internal or external application.

Direct arterial or venous pressure-monitoring modules with meter readouts also may be included.

The devices used should meet with the requirements and specifications of similar instruments used for other applications. However, special attention must be given to make the instrument as compact and with as few controls as possible, so that when pressed for time, operators do not make a mistake. Another important point requiring special attention is the power supply unit. A typical resuscitation trolley may have a power pack which consists of, say, a 70 ampere-hour storage battery, an invertor to convert the battery from 12 V DC output to 230 V AC and a battery charger. Switchgear and relays which automatically switch the trolley from the battery charge mode to self-contained power when the trolley is moved, to AC when the optional AC power cord is plugged into the mains, or to self-contained power when there is a line-power failure or the line cord is left unplugged, are also required.

Chapter 16

Patient Safety and Electromedical Equipment

In the past few decades, there has been an increasing quantum of electrical and medical electronic equipment used in our hospitals and health care delivery systems. If such equipment is in good working order, and installed properly, taking into consideration all the electrical safety standards and used correctly, it is relatively safe. The equipment itself is basically harmless, but because it makes conductive contact to the patient, it can be the key link in life-threatening current path. Occasionally, situations develop in hospitals where either the equipment becomes faulty or is used in a way that could precipitate an accident. Usually, the personnel using this equipment are not aware of the potential electrical hazards, which can be avoided with proper technical planning. This issue now lies in the attention of medical doctors and biomedical engineers. It is important that all biomedical engineers and hospital staff be aware of the electrical hazards likely to occur during patient monitoring and the safety precautions necessary to prevent them.

16.1 Physiological Effects of Electrical Currents

The table below summarizes the effects of a 50 Hz electric current on an average human body through the trunk (a one-second contact).

Current intensity 1 sec contact	*Effect*
1 mA	Threshold of perception
5 mA	Accepted as maximum harmless current intensity
10–20 mA	"Let go" current before sustained muscular contraction
50 mA	Pain, possible fainting, exhaustion, mechanical injury. Heart and respiratory functions continue
100–300 mA	Ventricular fibrillation will start, but respiratory centres remain intact
Greater than 6 A	Sustained myocardial contraction, followed by normal heart rhythm. Temporary respiratory paralysis. Burns are possible

Although it varies from person to person, the threshold of perception is about 1 mA. At this stage, a faint tingling sensation is felt. At 5 mA, many sensory nerves are stimulated and the sensation becomes painful and the subject jumps away from the source of stimulation. At the "let-go" current level around 10–20 mA, the muscle fibres are kept in a constantly depolarized condition. The subject then has no ability to control his own muscle actions, and he is unable to release his grip on the electrical conductor. At about 100 mA, other more life threatening physiological phenomena occur. One of the more common ones is ventricular fibrillation. At about 6 A or more, the heart goes into a sustained muscular contraction, but when the current ceases, the heart usually reverts to its normal co-ordinated mode of operation. Continuous high current levels of 6 A or more also cause temporary respiratory paralysis, and at levels much above 6 A, massive damage is caused by the heating effect of the current flow. These current values stated pertain to the net current flowing through the body.

The voltage required to obtain these current levels varies widely because the contact impedance varies from person to person, depending on the skin type, contact area, and whether or not the skin is moist. Typically, the threshold of feeling is from 10–50 volts.

5 mA has become accepted as the maximum current that should be allowed to pass through a human being from an external source. Therefore, 50 Hz leakage currents from the power line to an equipment case should be less than 5 mA. That is, if a person were to touch an electrical instrument while standing barefoot on wet earth, there should be no more than 5 mA flowing through his body.

16.2 Macroshock and Microshock

There are two modes of electric shock in hospitals: Macroshock and Microshock. *Macroshock* is an electric shock due to the contacts applied to the exterior of the body, the effects of which have been enumerated in the table above. Hospital patients may be exposed to macroshock from defective electric devices such as lamps, bed-adjusting motors and defective ECG machines in operation theatres.

Microshock is a much more serious hazard than macroshock. It applies to every patient who has a lead or electrical conductor from the interior of the heart, extending out through the body surface to pacemakers or cardiac catheters. Under these circumstances, there is a possibility that leakage currents might pass directly to the heart. In such conditions, a current as low as 50 μA caused by voltages as small as 5 mV can produce ventricular fibrillation of the human heart. For the patient's safety, it is now specified that any current passing internally through a patient should be limited to less than 10 μA, and that the non-therapeutic voltages that a patient may contact should be less than 5 mV.

Minimization of macroshock can be accomplished by the effective application of conventional safety methods, but the minimization of microshock requires a carefully designed and controlled electrical environment for the patient.

16.3 Preventive Measures to Reduce Shock Hazards

To protcct against shock, the two general techniques used are grounding and isolation.

Careful Grounding

To shield the patient from all sources of electrical shock, it is best to connect to one common point, all nearby metal surfaces and all electronic devices to which the patient is attached. This point, referred to as the Equipotential Patient Reference (EPR), is then connected to the hospital grounding system. This EPR can have potential with respect to the real ground, without any harm to the patient, because he cannot come to contact with any metal object except those tied to the EPR. It is important to limit the length of the ground wires connecting the equipment to the EPR, and to limit possible current flow in these wires so that the potential difference between the equipment and the EPR is always less than 5 mV.

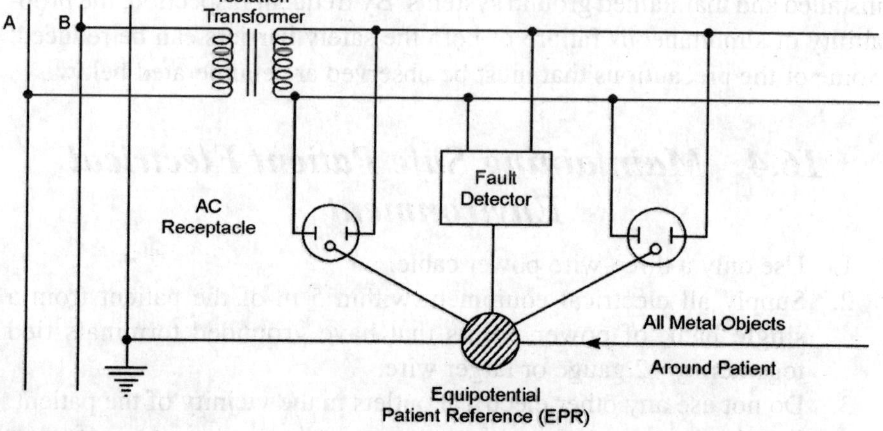

Figure 16.1 *Use of an Isolation Transformer and Equipotential Grounding*

Further improvements in an EPR system can be obtained with an isolating transformer, as shown in Figure 16.1. Either line A or line B can be short-circuited to ground without any current flowing through the 'short' connection. Here, even if a leakage path develops from the 'hotside' to a common point (EPR), the maximum current that can flow is that permitted by stray capacitance between the isolation transformer windings and in associated wiring. This system can also include a fault detector, as shown in Figure 16.1, that monitors the impedance between the EPR and each side of the line. If the impedance falls below a predetermined level, usually 60 000 ohms, that allows a few milliamps to flow in the EPR, a warning alarm is indicated.

Nevertheless, even with the improved protection provided by the EPR, not all hazards are eliminated. If the ground wire connecting the EPR to the chassis of an attached instrument should break, stray capacitance could raise the chassis potential above ground. Since power cords get stepped on and kicked around, breaks in ground wires occur far more often than leakage faults in transformers. If the installations are designed so that leakage current is less than 10 μA, the patient will be safe, should a ground wire break. Obviously instruments with isolated inputs provide an important safety factor, but many installations do not have them. Hence, good grounding practice is very essential and frequent inspection of the electrical environment of the patient monitoring system should be carried out.

From the above details, it can be seen that providing good and safe monitoring of patients in hospitals is not a simple task. The best that can be done is to reduce the risk of accidental electric shocks to the lowest possible level by using a combination of isolated patient circuits and a properly installed and maintained ground systems. By frequent inspection, the probability of simultaneous failure of both the safety barriers can be reduced. Some of the precautions that must be observed are enumerated below.

16.4 Maintaining Safe Patient Electrical Environment

1. Use only a three-wire power cable.
2. Supply all electrical equipment within 5 m of the patient from a single bank of power points that have grounded terminals tied together by 12 gauge or larger wire.
3. Do not use any other electrical outlets in the vicinity of the patient.
4. Supply the power points from a power line isolation transformer with ground-fault monitors.

5. Connect the conductive surfaces of all non-electrical devices within 5 m of the patient to a common ground terminal with separate 12 gauge, or larger, wire for each device. Connect the common ground terminal to the wall outlet ground.
6. Check the power cords for signs of damage and ground continuity—from the plug to the equipment case.
7. Use monitoring equipment with isolated inputs.
8. With all the equipment turned on, measure the potential between the common ground terminal and all conductive surfaces within 5 m of the patient. No surface should be more than 5 mV from the ground.
9. There should be periodic inspection of all the equipment and regular checks on ground continuity.

16.5 Earth-free Patient Monitoring

The use of single instruments with earthed patient circuits is not particularly hazardous. With the use of a number of devices connected simultaneously to the same patient, there are increased technical problems of protection. The growing use of intracardiac catheters and advances in cardiac surgery which provide for direct connection to the heart have greatly increased the electric shock risk to the patient. Therefore, the use of earth-free equipment is becoming imperative, except where it is technically impossible.

When patients are earthed, it causes many hazardous situations. Capacitors are usually connected between the mains supply and the earth chassis to eliminate mains-borne electrical interference. Such capacitors allow currents in the microampere range to flow between the live wire and earth. If the earth conection of an instrument is interrupted, the capacitative leakage currents can now flow through the patient, which can exceed the internal cardiac threshold. The effect of connecting other faulty equipment has already been discussed in the earlier paragraphs. With all the precautions taken with earthed-patient monitoring, safety is still dependent upon good maintenance and testing and a number of risks still remain, because the patient has an electrical path to the ground.

It is, therefore, important to eliminate all earth paths by designing instruments with earth-free patient circuits. An earth-free monitoring system is shown diagramatically in Figures 16.2 and 16.3. The system shown in Figure 16.2 has a mains transformer with an earthed interwinding screen to provide the major internal power supplies. Most of the circuits and all the exposed networks are earthed. Only a small portion of the circuit is connected directly to the patient or to a transducer which is in contact with the patient. This portion is well insulated from the earth and from all the

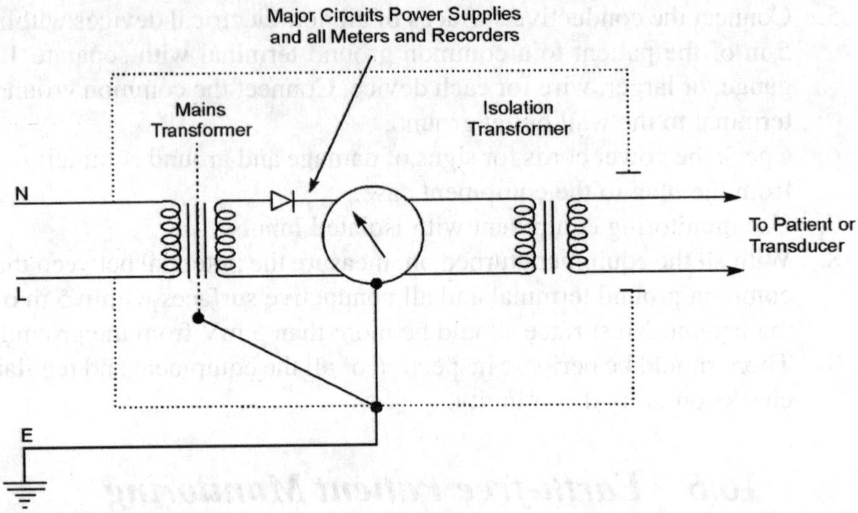

Figure 16.2 *A Basic Earth-free Instrument*

other circuits as well, to obtain complete shock protection. All the furniture and fittings associated with the patient should be as earth-free and antistatic as possible, so that the risk of shock on account of the passage of electricity to the earth through the body from a faulty device is minimal. When intracardiac devices are used or electrodes are attached to the heart, every reasonable step must be taken to keep these devices earth-free.

Thus, an earth-free system is much safer than the older system in which all the instruments and the patient were firmly earthed. Some of the electrical shock hazards during patient monitoring, discussed in this chapter, are attracting worldwide attention; and standard codes and safety regulations have been laid down and are being updated from time to time. The use of battery-operated monitoring and therapeutic devices would eliminate electric shock hazards. Till technology develops to this stage, one has to accept this risk and take all precautions to eliminate the lethal electrical hazards during patient monitoring.

16.6 Standards

On account of the major importance of electrical safety factors in the design, production and usage of medical equipment, standards have been set for this field. One such standard is given by the Bureau of Indian Standards under 'Medical Electrical Equipment – Part I: General Requirements for Safety IS 13450 (Part I): 1994'. The standards state that "Equipment shall be so designed that the risk of electrical shock in *normal use* and in

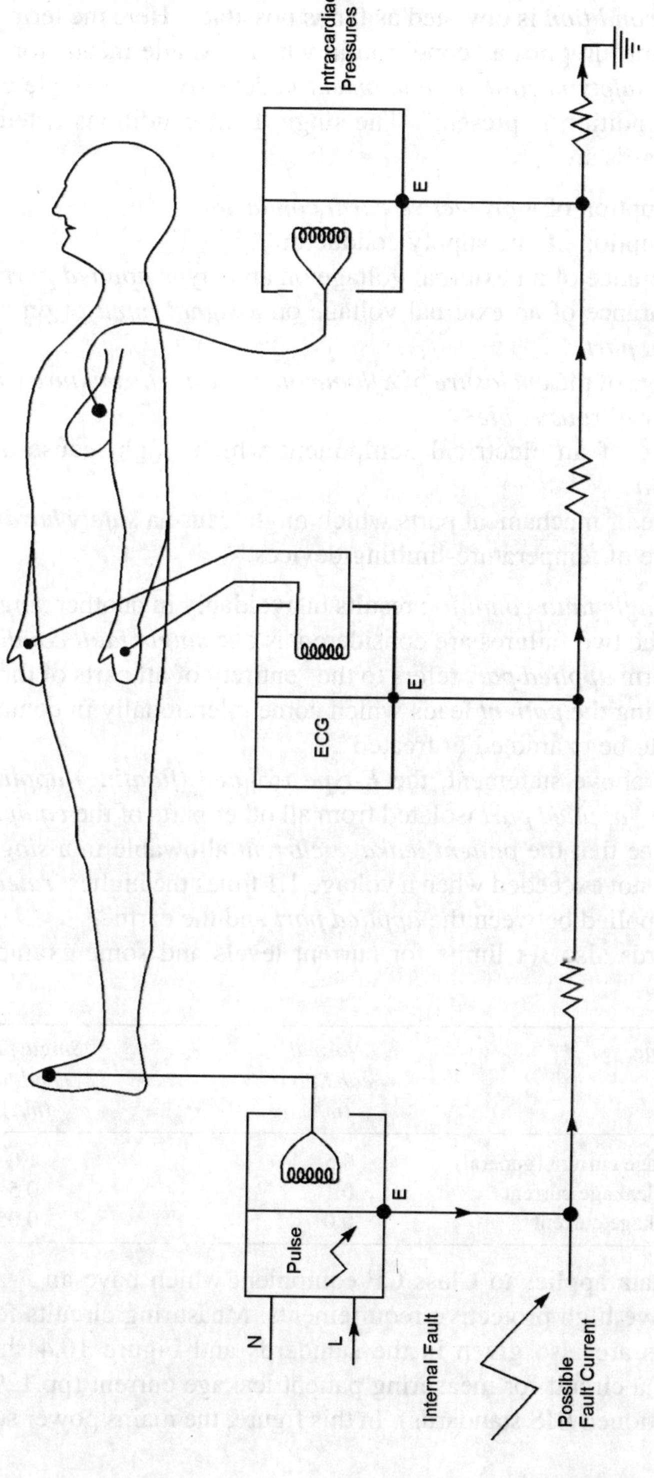

Figure 16.3 *A Earth-free Instrument Assembly, Free from most Risks*

single fault condition is obviated as far as possible". Here the term 'single fault condition' implies a "condition in which a single means for protection against *safety hazards in equipment* is defective or a single external abnormal condition is present". The single fault conditions listed in the above standards are:

a) interruption of a *protective earth conductor*
b) interruption of one supply conductor
c) appearance of an external voltage on an *F-type applied part*
d) appearance of an external voltage on a *signal input* or on a *signal output part*
e) leakage of the *enclosure* of a *flammable anesthetic mixture with oxygen or nitrous oxide*
f) failure of an electrical component which might cause a *safety hazard*
g) failure of mechanical parts which might cause a *safety hazard* and
h) failure of temperature-limiting devices.

"Where a *single fault condition* results unavoidably in another *single fault condition*, the two failures are considered as one *single fault condition*."

The term *applied part* refers to the "entirety of all parts of the *equipment* including the *patient* leads which come intentionally in contact with the *patient* to be examined or treated".

In the above statement, the *F-type isolated (floating) applied part* refers to the "*applied part* isolated from all other parts of the *equipment* to such a degree that the *patient leakage current* allowable in a *single fault condition* is not exceeded when a voltage 1.1 times the highest *rated mains voltage* is applied between the *applied part* and the earth".
The standards also set limits for current levels and some examples are given below.

Current type	*Normal condition (mA)*	*Single fault condition (mA)*
Earth leakage current (general)	0.5	1.0
Enclosure leakage current	0.1	0.5
Patient leakage current	0.01	0.05

The data applies to Class CF equipment which have an F-Applied Part and have high protective requirements. Measuring circuits for these assessments are also given in the standards and Figure 16.4 shows an example of a circuit for measuring patient leakage current (pp.129 of the above-mentioned BIS standards). In this figure, the mains power supply is

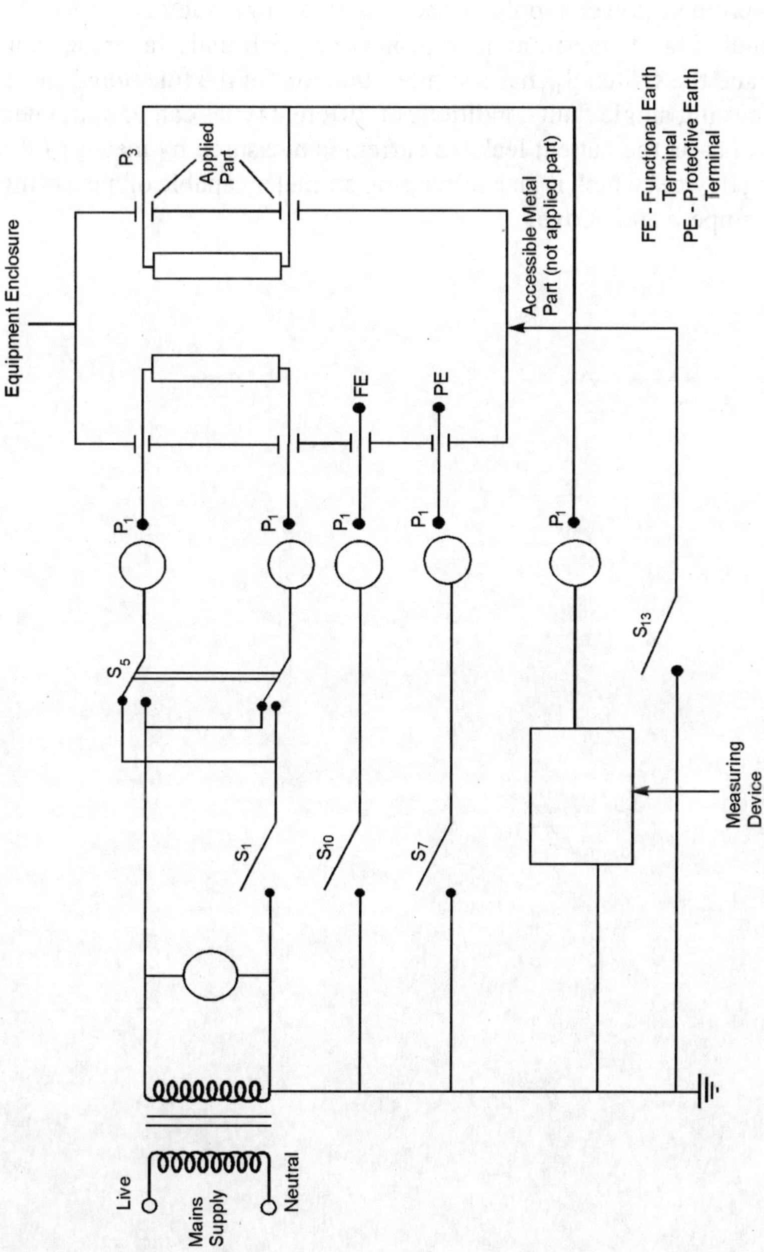

Figure 16.4 *The Measuring Circuit for Patient Leakage Current*

through a transformer and the switch S_5 allows for reversing the connection of the live and neutral terminals to the equipment. P_1 terminals are the sockets, plugs or terminals for the supply connection of the equipment. P_3 are the sockets, plugs or terminals for the patient connection as, for example, the leads of the ECG machine. S_1 is a single-throw switch used to simulate the interruption of power supply connection in a single fault condition. S_7 allows simulation of interruption of protective earth under a single fault condition and the switch S_{10} has a similar function for the functional earth. With this circuit, single fault conditions of different types can be simulated and, in each case, the patient leakage current is measured by means of the 'measuring device' which is an autoranging ammeter capable of measuring one microampere and above.

Chapter 17

Computer Applications in Medicine

Medicinc is one field where the whole range of computer power can be and, indeed in many cases, is being used to improve the standards of diagnosis and treatment of the disease without decreasing the need for the skills of highly-trained clinical and nursing staff. The range of applications is so great that it is possible to use only a few examples to illustrate what can be done. Eight key areas of computer applications in medicine can be identified:

i) Hospital Administration
ii) Clinical tests
iii) Laboratory tests
iv) Computer Assisted Medical Imaging
v) Computerized Aid to the Physically Handicapped
vi) Medical Diagnosis and Decision Making
vii) Patient Monitoring and
viii) Patient Management.

Figure 17.1 gives an overview of computer-assisted interventions in a hospital setting.

17.1 Hospital Administration

The computer can perform several 'on-line' and 'off-line' functions which facilitate the efficient handling of voluminous data and minimizes paperwork. A typical hospital information system covers the following application areas: reservations, admissions, discharges, transfers, outpatient registration, medical records index, master index (MPI), Soundex code, personnel registration, employee databank, patient billing and accounting, surgery and other procedure scheduling, pharmacy–patient medication, diet planning and cost–benefit analysis.

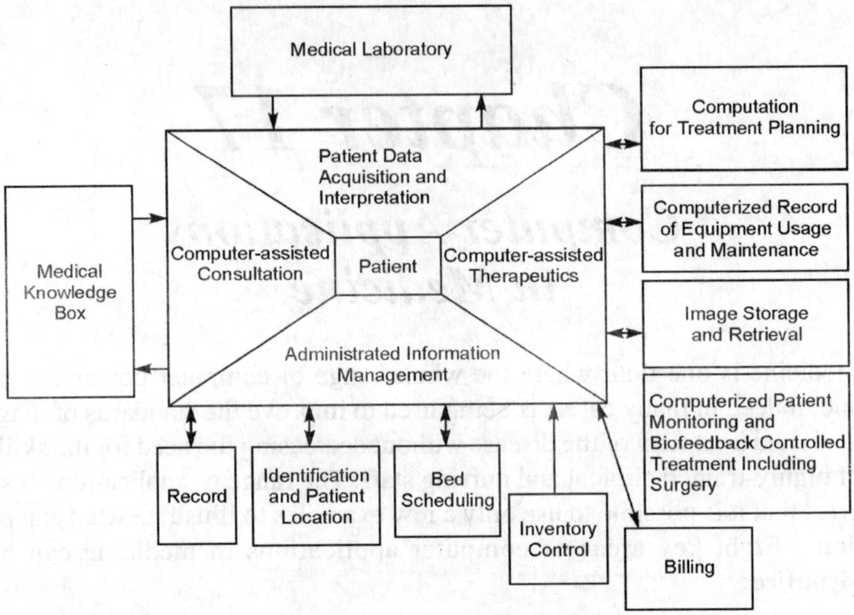

Figure 17.1 *Computer Assisted Interactions in a Hospital Setting*

17.2 Clinical Tests

ECG, EEG, and EMG have been subjected to automated analysis by computer. The analog signal from the ECG is converted by ADC into a digital signal. The signal can be represented as a (Fourier) series of sine and cosine waves, each of which has a frequency and amplitude term. The pattern-recognition algorithm of the computer can give an interpretation of the tracing, which is comparable to that of the expert cardiologist whose decision criteria have been incorporated into the program.

In a methodology developed by Dr. J.H.V. Bemmel* for automated classification of ECG into normal and abnormal categories, the R-wave of the ECG is first detected. Most ECG interpretative systems are based upon three simultaneous lead ECGs. A detection function, d(i), where (i) is the sample number, has been defined in terms of three-dimensional vector cardiography. The ECG is expressed as

$$\text{ECG}(i) = \left[X_1(i), X_2(i), X_3(i)\right] \quad ...(17.1)$$

* Bemmel, J.H.V (1985): *Recognition of Electrocardiographic Patterns in Medical Informatics*, Amsterdam – 10 Year Review, pp 21–46.

where $X_1(i)$, $X_2(i)$, and $X_3(i)$ are the ith sample values of the ECG in the three leads. Then, the detection function becomes

$$d(i) = X_1^2(i) + X_2^2(i) + X_3^2(i) \tag{17.2}$$

So, an ECG wave is considered to be a true QRS complex, if it fulfills the following criteria:

> d(i) > 25% during > 10 ms and some d(i) > 40% during 180 ms thereafter and the distance to a preceding QRS complex is greater than 250 ms.

After the QRS complex is detected, it is matched with stored normal and abnormal waveforms in a template matching format to arrive at a classification. The number of computational steps is large and so the data processing is done with the help of a computer.

In EEG analysis, particularly in an evoked potential, the particular capabilities of the computer that are useful with respect to frequency analysis, are correlation analysis (auto-correlation, cross-correlation) and averaging.

17.3 Laboratory Tests

The utilization of the autoanalyzer capable of measuring the concentration of 18 or more chemical substances in a single sample has provided a large database, which is handled competently by the computer for appropriate interpretation and multivariate analysis.

A laboratory system is an interactive data management system. Computerization increases the throughput of laboratory tests, reduces transcription errors, ensures quality control, rapid performance and dissemination of test results.

The gas chromatograph mass spectrometer (GCMS) is the most powerful analytical instrument used in the clinical laboratory. High sensitivity built-in computers allow on-line recording of test results, which help in clinical decision making.

Chromosome analysis and cytological diagnosis can be achieved by the technique of syntax-directed pattern recognition.

17.4 Computer-assisted Medical Imaging

Nuclear Medicine was the first clinical discipline to use computers in medical imaging; wherein the computer is used to supplement human performance, to store large amounts of data, quantify them and carry out memory and quantification functions that cannot be accomplished by the human brain alone.

New concepts and developments in computer science and applied mathematics have made possible Computer-assisted X-ray Tomography (CT), which has made a phenomenal impact on the practice of medicine. The computed CT consists of two sequential processes. First, the absorption data of the beam of X-rays within a single cross-sectional plane must be acquired. Second, those data must be used as input to a computer which calculates about 90 000 attenuation coefficients, and achieves image reconstruction by a mathematical technique called 'filtered back-projection'. The fourth generation of CT scanners can achieve this in one second, in a one millimeter slice of the human body, with a spatial resolution of 1–2 mm on a scale of 2 000 levels of attenuation (–1000 for air, +1000 for bone, 0 for water) and tissue density differences can be detected. Further details of this technology are given in Chapter 12.

Computerized emission tomography, either with a single photon (SPECT) or positron (PET) uses the same mathematical technique of filtered back-projection to reconstruct an image of distribution of radioactivity in an organ. Quantitative regional physiology and biochemistry (blood flow, oxygen consumption, glucose consumption metabolism, neurotransmitter receptor mapping) is now possible in the living human brain and it is also possible to relate these changes to mental and behavioural activity. Diagnosis of mental disorders such as schizophrenia, manic-depressive psychosis and Alzheimer's dementia, is now objectively possible.

NMR imaging (neugmatography) is based on the conversion of proton density and $T_1 + T_2$ relaxation signals by Fourier transformation and Radon transformation into images. Electronically, it is possible to achieve image reconstruction in multiple planes (such as cross-sectional, sagittal, and coronal). NMR images provide greater contrast between the grey and white matter of the brain compared to the X-ray or CT. It gives information about tissue chemistry in the living body, which will help in the differentiation of a normal tissue from cancerous tissue and even precancerous tissue.

Also, the standard X-ray images can be processed by a computer algorithm to enhance the contrast and display pathology clearly. One such processing algorithm applies a 'high-pass filter mask', according to the expression

$$F_{(m,n)} = \sum_{y=-1}^{1} \sum_{x=-1}^{1} O_{(m+x,n+y)} \cdot H_{(x+z,y+z)} \qquad \text{....(17.3)}$$

where

$$H = \begin{bmatrix} -1 & -1 & -1 \\ -1 & 9 & -1 \\ -1 & -1 & -1 \end{bmatrix} \qquad ....(17.4)$$

$F_{(m,n)}$ = the filtered image made up of $m \times n$ image points,
$O_{(m,n)}$ = the original image, and
$H_{(x,y)}$ = the high-pass filter mask.

17.5 Computerized Aid to the Physically Handicapped

Computer analysis of accelerations and rotations of the limbs and trunk, and finite element analysis of bone stresses, enables prostheses to be designed with the necessary strength and functionality.

For the Blind: The Optacon is a direct translation device that converts an optical image of a printed character on a page of a book or a newspaper to a vibrating, tactile facsimile. Electrical output signals from the sensor array are processed in a simple computer system. A blind reader with his fingertips on a tactile pad can feel the tactile facsimile, like the Braille symbols.

Graphics can also be communicated to the blind students by means of a computer. Figure 17.2(a) shows a classroom situation where there is a computer on the teacher's desk and the monitor is hooked to several monitors one each on every student's desk. If a diagram is fed into the computer at the teacher's desk, the diagram is displayed on both teacher's and students' desk monitors. Each student is provided with an optical sensor linked to a vibrator, as given in Figure 17.2(b). When the blind student scrolls the optical sensor over the monitor screen, a tactile vibration is felt at illuminated point on the monitor screen. A globule of modelling clay is stuck onto the monitor screen at each point where a vibration is felt. Thus, the diagram is traced with globules of modelling clay. The blind student can then 'feel' the diagram's shape by passing his/her fingers over the globules. Later on, the globules may be removed, by wiping them off, to make the monitor ready for reading another diagram.

Voice Synthesis: Microprocessors help the aphasic or dysarthric patient to generate simple commands, as in shopping. A portable computer carries the necessary vocabulary.

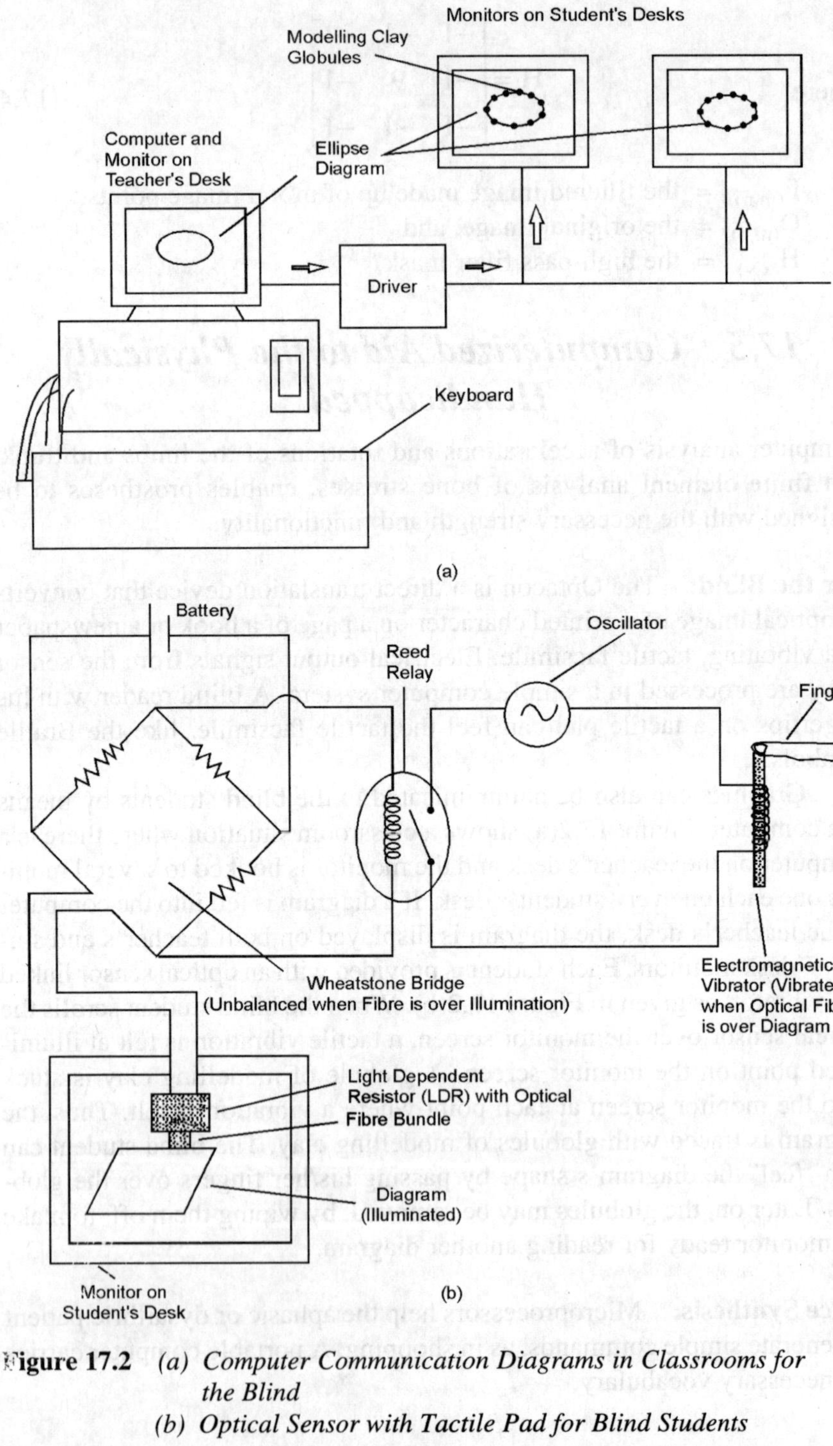

Figure 17.2 *(a) Computer Communication Diagrams in Classrooms for the Blind*
(b) Optical Sensor with Tactile Pad for Blind Students

For the Deaf: An auditory prosthesis is now available, based on electrical stimulation of the auditory nerve, in cases where the transduction machinery of the cochlea is totally defective.

For Quadriplegics and Paraplegics: Manoeuverable, lightweight, portable wheelchairs are available, wherein microprocessors control the speed of movement. Prosthetic limbs have been developed with pre-programmed or interactive computer algorithms which obey the user's voice-commands like "wrist, arm, grip, raise, lower, right, left, in, out, off".

17.6 Medical Diagnosis and Decision Making

Computers are used in clinical decision making to improve diagnostic accuracy by using systematic approaches and explicit decision criteria; to enhance diagnostic reliability by precluding unwarranted influences of non-identical but similar cases; to increase cost efficiency; to advance understanding of the structure of medical knowledge; and to gain insight into clinical decision making itself.

Two streams of research on medical decision making can be identified—one stream consists of the development of mathematical tools with which the physician may sharpen his decisions (Decision Trees, Bayes' Theorem, Decision Matrices, ROC analysis, Cost-Benefit relationships) with the purpose of maximizing the expected utility of his decisions.

In the Bayesian approach, with the following notations:

- n – number of possible diseases causing symptons
- D_i – ith disease
- P(D) – *a priori* probability of occurrence of disease D_i
- S – set of facts about a particular patient, that is, signs and symptoms like swellings and headaches
- $P(S/D_i)$ – conditional probability of S given the presence of D_i, that is, how often D_i would result in specific signs and symptoms designated by S,

the *a posteriori* probability is

$$P(D_i/S) = \frac{P(D_i) \times P(S/D_i)}{\sum_{j=1}^{n} P(D_j) \times P(S/D_j)} \quad \text{....(17.5)}$$

where $P(D_i/S)$ is the probability of the disease D_i, being the cause of the sign and symptom set S. The computations require computers on account of the larger number of steps.

The other stream consists of studies of how physicians make decisions. These studies are related to artificial intelligence (AI) and they emphasize computer augmentation of clinical judgment.

Another emerging direction in computer-aided decision making combines computer graphics with medical system analysis. *Computerized cephalometrics*, which implies a study of the dimensions of the head region, is an example. There are important applications of the methodology in dentistry to help in assessing maladjustment of teeth and provide guidance in misalignment correction in the orthodontic speciality of dentistry.

A system named DIG-CEPH has been developed at Indian Institute of Technology (IIT), and the All India Institute of Medical Sciences (AIIMS), New Delhi. A X-ray of the head region of the patient is placed on an illuminated screen of a digitizer. The digitizer is connected to a personal computer. When the doctor places the mouse of the digitizer on a specific landmark of the head as seen on the X-ray, as for example the meeting point of the vertical part and the horizontal part of the jaw bone, and a button is pressed, the co-ordinates of the landmark are automatically entered into the computer. From the co-ordinates, a computer program using the Turbo, Pascal language constructs a graphic image as shown in Figure 17.3; also from the co-ordinate data a number of parameters relating to angles are computed. The program also carries out analysis to provide information to help the dentist to apply appropriate forces using stainless steel wires to correct teeth positions. Figure 17.3 also shows the position of the teeth before correction and after computer-assisted correction. With such a correction, the facial features improve and mastication becomes more effective.

The Future: "It seems probable that in the not-too-distant future the physician and the computer will engage in frequent dialogue, the computer continuously taking note of the history, physical findings, laboratory data, and the like, alerting the physician to the most likely diagnosis, and suggesting the appropriate, safest course of action" (from: W.B. Schwartz, *Medicine & Computer* (J) (1982) **20**: 357).

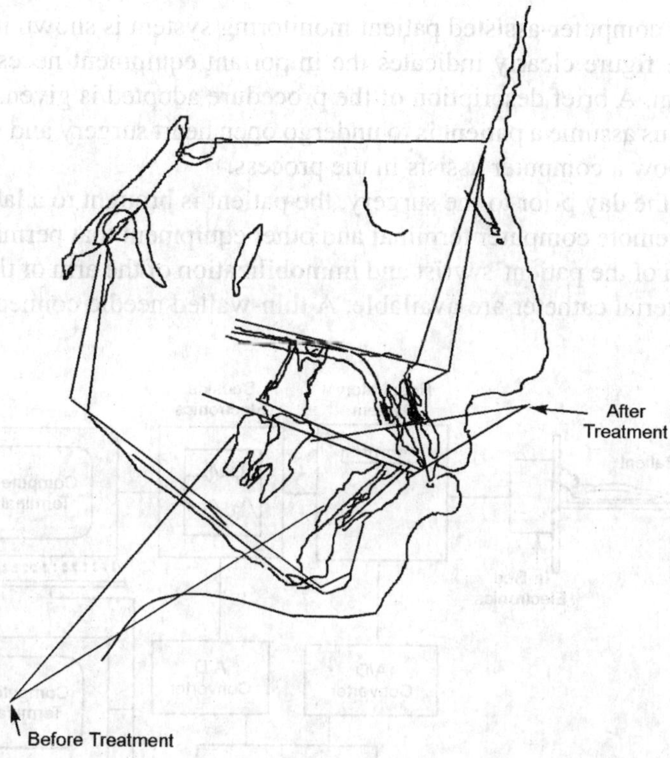

Figure 17.3 *Digital Cephalographic Image of "Before and After" Treatment for Malpositioned Teeth (11-year-old Female)*

17.7 Patient Monitoring

a. Patient Monitoring in ICU and Automated Therapy

In the case of a critically ill patient, all the important physiological parameters (temperature, pulse rate, respiration rate, blood pressure, central venous pressure, heart function, lung function, urine output and blood chemistry) are transduced into electrical signals and converted through AC and DC into digital signals which are stored and analyzed by the computer. Alarm signals are generated according to predetermined parameters.

The ultimate use of computers in patient care is in a closed control system, where the computer not only monitors the patient's state but is also equipped with effectors to correct any deviations from normal values.

b. Computer-assisted Patient Monitoring System

A typical computer-assisted patient monitoring system is shown in Figure 17.4. The figure clearly indicates the important equipment necessary for the system. A brief description of the procedure adopted is given.

Let us assume a patient is to undergo open heart surgery and see step-by-step how a computer assists in the process.

On the day prior to the surgery, the patient is brought to a laboratory where a remote computer terminal and other equipment that permit hyper-extension of the patient's wrist and immobilization of the arm of the monitoring arterial catheter are available. A thin-walled needle connected to a

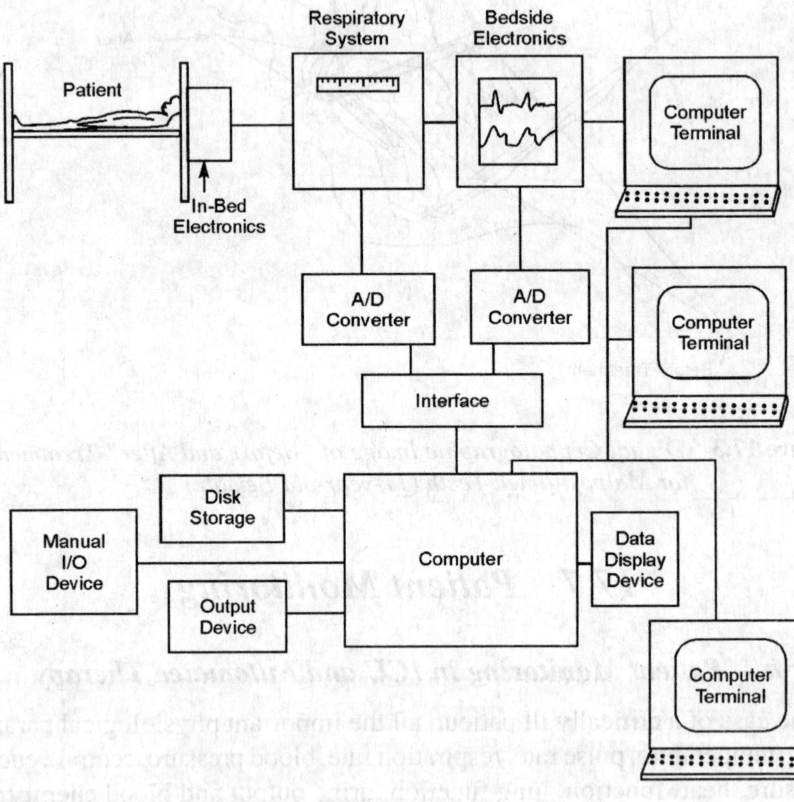

Figure 17.4 *A Typical Computer-assisted Patient Monitoring System*

strain gauge pressure-transducer, by way of a special catheter filled with saline solution, is inserted into the radial artery. The pressure being transmitted through the needle is visualized on an oscilloscope as the needle is advanced, until the tip lies in the subclavian artery. Now, control measurements are initiated. A cardiac output value determined earlier, at a prior diagnostic heart catheterization, is entered into the computer. Calculations of the required parameters are then made.

When the patient is wheeled out for surgery, the catheter is connected to a suitable strain gauge transducer which is strapped to the patient's forearm. Using a remote terminal of the computer at the operation table, the anesthesiologist selects the calibration option. He gets from the computer, the stroke volume, heart-rate, cardiac output, mean pressure, systolic pressure, and other parameters as the surgery proceeds.

On the completion of the operation, the patient is taken to the intensive care ward. The following information is directly sent to the computer through suitable transducers:

i. Central arterial pressure wave
ii. Central venous pressure wave
iii. Electrocardiogram.

The data collected by the anesthesiologist during surgery are checked on the computer.

When the doctor presses the CALL button on the computer terminal and feeds the patient's number, the computer displays the patient's name, age and a list of other details pertaining to the patient. For each heart cycle, the computer determines the following variables: stroke volume, heart-rate, cardiac output, duration of systole, peripheral resistance, systolic pressure, respiratory rate, respiratory amplitude, heart-rate from ECG, *etc*. These give the baseline measurement. Subsequent measurements given by the computer will be compared with the baseline measurements to determine whether there is a statistically significant change in the patient's status. Whenever a deviation occurs a, red light and print-out will be given by the computer calling the attention of the doctor to the fact that a change of status has occurred. The schedule of measurements is done under the control of the computer.

The computer also provides a number of important options to the doctor. These options are:

i. The doctor, on demand, can obtain a measurement of the physiological state of the patient.
ii. The doctor can ask the computer to plot the time course of one or more variables over a required time period.
iii. The doctor can ask for plots of central arterial pressure recorded at the last measurement, superimposed on the base measurements.
vi. Instantaneous information on the fluid intake and output status of the patient may be seen on the display and reviewed.

The computer also provides the schedule of activities to be performed by the nurse. It warns the nurse what activity is overdue. There will be a data display device which is used for the display of reports by the computer and for interactive communication between the computer and the medical staff.

17.8 Patient Management

Computers have been widely applied to the problems of patient management and record-keeping in hospitals.

Development in word-processing equipment, with user-friendly features for modification of data, insertion or deletion; the exploitation of the tree structure of viewed data which gives access to substantial databases using simpler single-key instructions; special keyboards allowing easy entry of keywords or the use of light pens, tracker balls or touch wires to select from a 'menu'; and, in the longer term, developments in voice and optical character recognition, can all lead to computer-based systems which are more acceptable to the medical profession.

Another area of patient management where computer-based systems are proving to be of value is in research. One such system, based on the use of the ANSI standards, is MUMPS. A detailed file is created which includes information about biochemistry, radiotherapy, haematology, histopathology, nuclear medicine, pharmacology, diagnostic radiology, surgery and a general clinical file; the data being entered via a standard, general video display unit.

MUMPS (Massachusetts General Hospital Utility Multi-Programming System)

It is specially designed to meet medical requirements. MUMPS differs from structured languages such as PASCAL and FORTRAN and COBOL that emphasize the delineation of data sizes and types to restrict data definition in favour of control. These three languages presume that the problems that may arise can be anticipated in advance. In medicine, this is seldom the case. MUMPS accepts definitional ambiguities at programming time and resolves the issue at the time of actual use, relying on special routines to screen for errors and to establish control. The development of LISP (List Processing) and PROLOG (Programmed Logic), the two favoured languages for (AI) Artificial Intelligence, was a major breakthough in the field of non-numeric data processing. Based on LISP, many super high-level languages have been created. Newer tools like the predictive calculus, frame systems, rewriting rules, semantic networks and production rules, may be characterized by the term "propositional representations". The notion of AI is defined as the development and application of propositional representations.

It is possible to collate data about diagnosis, therapy, actiology and prognosis for any specified required group of patients. Analysis of patient data is now becoming possible, which was previously ruled out by the sheer amount of work involved in extracting information from notes which were often incomplete, unintelligible or even missing.

What has been illustrated is one example of computer-based analysis of the results of various biomedical measurements.

Normally, in big hospitals, a central powerful computer will be connected to a number of small computers or intelligent terminals. This will provide some failure-proof arrangement.

The computer will have enormous storage facility and real-time data processing capability. The best computer for medical applications will also have a time-sharing facility.

Chapter 18

The Hospital and the Medical Electronics Department

Medicine today relies heavily on electromedical equipment. Even though much effort is made to design and build it to high standards, such equipment is not foolproof and must be used properly. With the increasing life-saving value of equipment comes its risk of injuring a patient or even a hospital personnel, if it is misused or becomes faulty.

18.1 Medical Responsibility

For the patient, electromedical equipment now means treating what was previously untreatable. But to the physician and nurse, it means knowing more about the devices at their disposal. And this means being able to use equipment properly. Medical personnel can now treat a more seriously ill person, but because of this increased seriousness, the devices at their disposal have become correspondingly more critical in value.

Medical personnel – the physician, the nurse, the technician – now bear the responsibility for knowing how to use the equipment. Little or no training in the use of equipment can no longer be tolerated in the light of the effects such equipment can have on patients; and the unknowing misuse of equipment can be a very costly mistake.

18.2 Using Equipment

As with any endeavour, proficiency demands study and practice. As physicians and nurses study the science of medicine, a portion of that learning time must now include the use of equipment. The steps in learning to use electromedical equipment include: the study and understanding of basic principles of electricity before and during operation of the equipment, referring to the 'User Manual' relating to the equipment supplied by the manufacturer (taking great pains and written in simple terms), and spending some prescribed time or orientation period in the actual use of the equipment (especially major or complicated versions).

Whenever equipment is to be used, several checks should always be made.

- A visual examination of equipment (say, for missing knobs) can reveal its status to some extent.
- A manual examination reveals much more about the condition of equipment, *e.g.*, condition of power cords or patient cables.
- An operational examination, with regard to proper power supply, condition of controls or alarms, *etc*.
- In-use examination, *i.e.*, after the pre-patient checking, a check of the equipment after it is attached to the patient is to be made. This is when operational abnormalities may be discovered. After-use examination, *i.e.*, after using the equipment, some thought should be given to removing it from an area and retaining the power cords, *etc.*, in a proper condition.

These four steps should become standard and routine every time any equipment is used, involving just a few well-spent minutes.

18.3 Design of Electromedical Equipment

The major design problem with electromedical equipment is its use by non-technical persons. Just as the car or television is designed for a predominantly non-technical public, so, too, most electromedical equipment must be designed for a predominantly non-technical public. Of almost equal consideration and importance is the fact that electromedical equipment is *critical*. Some may mean the difference between life and death. Because of this reliance by the medical community, the design of such equipment requires some extra effort in order to meet this expectancy.

In the design of equipment, a manufacturer evaluates many parameters in determining the components that are finally chosen. These include the following:

1. How long equipment will be continuously used at any one stretch, *e.g.*, 1 hour/1 day/1 week
2. How accurate the device must be in order to be useful, *e.g.*, 1, 2, 5%
3. Under what conditions the equipment will be used and to what stresses it will be subjected
4. How life-saving in value the equipment is, *i.e.*, what a failure to operate would mean
5. What financial consideration must be taken into account, *e.g.*, will the final cost mean only a few hospitals can afford the device?

These factors must all be weighed in determining the final equipment design. They subsequently determine the operational limits of the equipment. Medical personnel should be aware, through reading the user manual, of the operational limits of the device.

18.4 Technical Responsibility

A growing number of hospitals are realizing that the proliferation of electromedical equipment has reached a level warranting the hospital staffing qualified technical personnel. In addition, medical electronics laboratories (MELS) are being established to enable these individuals to utilize their skills. The following discussion is based on a hospital having such trained persons on their staff.

At one level is the Medical Electronics Technician (MET) who is trained to repair electromedical equipment. Beyond the repair perimeter, are other problems which require the services of a Medical Electronics Engineer (MEE). The roles of the MEE and MET are discussed in the subsequent sections.

18.4.1 Servicing Priorities and Conditions

When a call is received in the MEL reporting a faulty piece of equipment, its repair is accomplished based on a set of priorities. If a back-up unit is available, a technician can bring it to the area needing the equipment and remove the faulty equipment. If no back-up is available and its repair is vital, a technician will, of course, immediately attend to it. Priority is normally established for the operating room, special care units, and catheterization laboratories. No hard and fast rules can or should be made. All patient equipment is vital.

The operating room presents the most unusual conditions in the repair of the equipment. If the equipment cannot be removed from the operating room, the technician must be prepared to don an operating-gown and repair the equipment. In some instances, this may occur during an operation. This requires a knowledge of repairing the equipment following sterile operating room procedures.

Figure 18.1 is a pictorial representation of the maintenance cycle for electromedical equipment—from discovery of problems by the medical personnel, to the repair of the equipment by qualified personnel, to the return to use in good working order.

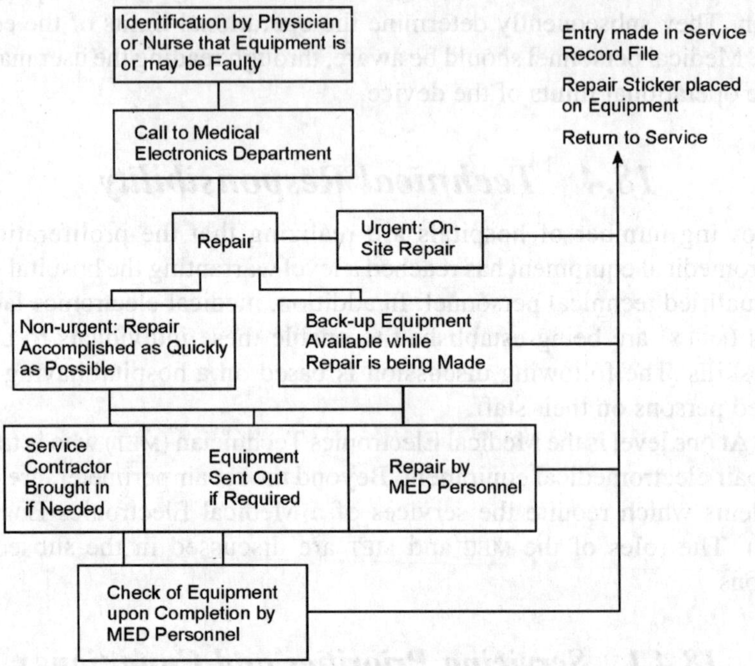

Figure 18.1 *Maintenance Cycle for Electromedical Equipment*

18.4.2 Preventive Maintenance

Preventive maintenance (PM) on equipment is one of the most effective ways of reducing the number of equipment failures and increasing its useful life. By periodically checking the equipment, minor problems can be corrected before they become major problems. Since each piece of equipment has its own schedule for PM, a hospital with hundreds of pieces of equipment must develop a workable schedule so that all equipment is inspected and checked with regularity. A PM program should be initiated on a small scale and gradually expanded as experience is gained.

18.5 Administrative Responsibility

Administrative responsibility for maintaining the equipment extends to employing qualified persons to maintain the equipment and providing funding, so that these persons can adequately perform this function. Hospital administrators should realize that the use of electrical equipment has proliferated; and, to continue to provide patients with a high degree of safety, it is now essential to have personnel trained in electronics on hand at all times to maintain the equipment.

Additionally, these medical electronic personnel can act as advisors to an administrator on equipment purchases and problems.

18.6 Hospital Responsibilities

The hospital shares a large responsibility in determining the direction electromedical equipment, in general, will take in the future. The condition is borne out by the role hospitals have in treating patients. Medical personnel must meet certain standards. Hospital regulations exist for dispensing drugs, sterile procedures, operating room standards, *etc*. Similarly, the *hospital* bears the prime responsibility for safe utilization of electromedical equipment.

This responsibility extends to three areas in which the hospital must be able to operate effectively. They are:

a. Meeting in-house technical needs
b. Communicating effectively to the manufacturing community the equipment requirements of the hospital
c. Establishing standards for equipment.

The solution to meeting these responsibilities is by the establishment of a Medical Electronics Department within the hospital (discussed below). The solution to (a) through (c) involves having qualified in-house personnel to speak for the hospital.

18.7 The Medical Electronics Department

The hospital has become a big operation; the medical services now performed include operations lasting over 6 hours, requiring several surgeons and many nurses, utilizing equipment both costly and invaluable, and requiring a long and expensive recuperation period. As part of this world of treatment, the role of electronics is and will continue to be one of serving medicine by producing devices which meet medical needs.

To achieve equipment order, a Medical Electronics Department (MED) can be established. Its activities can be as wide as the equipment it services (See Figure 18.2).

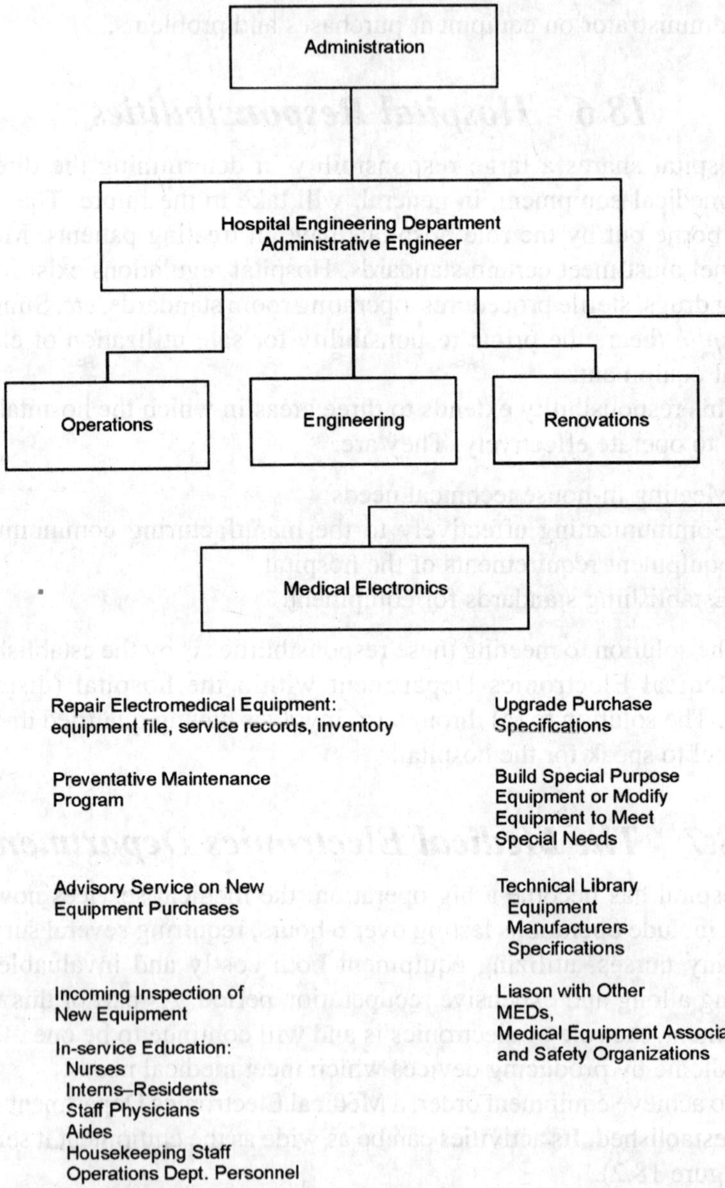

Figure 18.2 *The Medical Electronic Department within the Hospital*

18.7.1 One MED

It should be evident that because the use of electronics has spread throughout the spectrum of medical services, the MED must be its own entity and not be part of one particular medical service. Further, there should be only one MED which is responsible for all the electromedical equipment in the hospital. The MED can, thus, become the resource centre for all the needs of medicine as it relates to electromedical equipment. As a result, the knowledge gained from the equipment of one department can be used in other departments.

18.7.2 The MED within the Engineering Department

The MED should be established within the framework of a hospital's engineering department. Figure 18.2 includes a block diagram of how this might occur. This diagram will vary, depending on the size of a hospital and the scope of its medical activities.

There are several reasons for this placement within the engineering department:

1. Medical electronics is a physical type of service, providing medical staff with devices for them to use in their functioning.
2. Medical electronics is basically a technical function, solving medical problems through engineering and the use of electronic equipment.
3. For the staff within the MED, most communication is in technical terms. After an equipment problem is defined (*e.g.*, the monitor scope does not display the ECG pattern), the solution is accomplished in technical terms (*e.g.*, replacement of faulty components).

After analyzing points 1 through 3 we can see that the Engineering Department is the logical department in which MED can best function.

Figure 18.2 shows the activities which the MED is capable of handling. How close the MED comes to performing all of these functions is a result of two factors—the staffing of the department and administrative support.

A difficult part of establishing the MED is defining how much money is required for its operation. The salary of the persons hired, the cost of laboratory equipment, overhead expenses—all these costs can be estimated. But other cost factors are involved and must also be considered. If no central equipment service records are maintained, a review of the accounting department's records can provide cost data on labour and parts charges by service contractors.

To justify establishing a MED on a purely cost basis, is not looking at the situation in its entirety. The increased benefits of having an in-house capability must be considered for such a study to be realistic.

18.8 Resident Technical Personnel

We have stated the need to staff the MED with qualified personnel. Given below are the descriptions of the persons. A hospital's commitment to an electromedical equipment program will determine the size of the staff.

18.8.1 The Medical Electronics Technician

The person most directly involved with the equipment and its actual use by nurses and physicians is the Medical Electronics Technician (MET). He must be capable of not only repairing electromedical equipment, but also knowing how to use it. In servicing such medical equipment, there will be times when the equipment fails and the MET will be required to evaluate the problem on-site, (*e.g.*, the operating room or the ICU). He will, thus, need a background in physiology to adequately solve the problem. Also, to properly repair and maintain the equipment, he will need to know the conditions under which it is used; again, direct interaction with medical personnel becomes necessary.

Additionally, the MET will need to be extensively knowledgeable in electronic laboratory equipment. For it is his skill with such equipment, in addition to a background in electronics, that will determine how well the equipment is maintained.

18.8.2 The Medical Electronics Engineer

A Medical Electronics Engineer (MEE) will be involved with a wider variety of problems within the medical equipment field. He will interact not only with medical personnel, but also with the manufacturing community. In the hospital he will: (a) plan for preventive maintenance programs; (b) determine, along with medical and administrative personnel, the equipment needs of the hospital to maximize utilization and minimize duplication and confusion; (c) evaluate new equipment for integrity; (d) provide the hospital's in-service education department with lectures and demonstrations; (e) investigate accidents involving equipment; and (f) work with medical personnel on new features for redesigning equipment to meet a new purpose. For the hospital, he must keep abreast of all the existing equipment and its condition, as well as supervise its maintenance through the MET. He also needs to keep abreast of the development of any new equipment.

The MEE will also be the hospital's technical adviser, with regard to the manufacturing community. In meetings with manufacturers, he will advise the hospital on the company's technical capabilities. He will also administer the hospital's minimum performance standards for equipment, by working with his technical counterpart in the industry.

18.8.3 The Technical Administrator

The need for *technical administrators* is growing because of the effect electromedical equipment has on the treatment of the patient. Standards that assure a high degree of safety during equipment use must involve all hospitals. Equipment policies mean a combined agreement of the *medical groups* that use it, the *manufacturing community* that will build it, and the *government* which has the obligation of protecting the rights of the public.

A technical administrator for electromedical equipment would set hospital policies for equipment. He would interact extensively with state and central committees, with medical and engineering societies, with "standards" committees, and with manufacturing associations. He would communicate the hospital's needs and requirements to each group.

These three persons are professionals in their area of expertise. As such, liaison with their counterparts in other hospitals is essential for themselves, for their field, and for their hospital. Their participation at conferences and attendance at conventions is very essential. Also, continuing education is a necessity to keep abreast with advances in technology.

18.9 First and Last, the Patient

The question of safety when the practice of medicine is involved is, of concern to everyone—the physician, the nurse, the technician, the engineer, the administrator. For those in direct contact with patients, the concern is self-evident. When the distance from the patient increases, the personal concern decreases and other valid concerns increase. Examples include proper scheduling to ensure adequate medical coverage at all times; establishing the best system for distributing food; determining a cleaning schedule to ensure that bacteria level is kept to a minimum; maintaining an adequate supply of drugs; assuring emergency power supply when needed and to most critical areas first; and keeping the building safe from unwanted visitors. The heavy influx of electronics adds another dimension to that concern.

So, it is now the task of the medical community to meet the responsibilities that accompany the use of electronics. Thus, Medical Electronics has become a significant factor in the total operation of the hospital's basic role of treating the patients. The highest degree of safety must be assured for the person who will be on the receiving end of this electronics—the *patient*.

Bibliography

Arumugam, M. (1992) *Biomedical Instrumentation*, Anuradha Publishers, Kumbakonam, India.

Aston, R. (1990) *Principles of Biomedical Instrumentation and Measurements,* Merrill Publishing Company, Ohio, USA.

Basmajian, J.V. (1980) *Grant's Method of Anatomy,* Williams & Wilkins, Baltimore, USA.

Bronzino, J.D. (1977) *Technology for Patient Care*, The C.V. Mosby Company, St. Louis, USA.

Bronzino, J.D. (Ed.) (2000) *The Biomedical Engineering Handbook,* (Vol. 1 and Vol. 2) CRC Press, Boca Raton, USA.

Brown, J.H.U., Jacobs, J.E., and Stark, L. (1971) *Biomedical Engineering*, F.A. Davis Company, Philadelphia.

Bushberg, J.Y., Seibert, J.A., Leidholdt, E.M. and Boone, J.M. (1994) *The Essential Physics of Medical Imaging*, Williams & Wilkins, Baltimore.

Cobbold, R.S.C. (1974) *Transducers for Biomedical Instruments. Principles and Applications*, John Wiley and Sons, New York.

Cromwell, L., Weibell, F. and Pfeiffer, E.A. (1993) *Biomedical Instrumentation and Measurements*, Prentice-Hall of India Pvt. Ltd., New Delhi.

Ganong, W.F. (1995) *Review of Medical Physiology*, Lange Medical Books (Prentice-Hall International Inc.), Connecticut, USA.

Geddes, L.A. (1972) *Electrodes and the Measurement of Bioelectric Events*, John Wiley and Sons, New York, USA.

Geddes, L.A. and Baker, L.F. (1975) *Principles of Applied Biomedical Instrumentation*, John Wiley and Sons, New York, USA.

Guha, S.K. (1969) *Introduction to Medical Electronics*, Bharati Bhavan, Patna, India.

Guha, S.K. (1990) *Bioengineering in Reproductive Medicine*, CRC Press, Boston, USA.

Jacobson, B. and Webster, J.G. (1977) *Medicine and Clinical Engineering*, Prentice-Hall Inc., Englewood Cliffs, USA.

Jones, L.D. and Chin, A.F. (1983) *Electronic Instruments and Measurements*, John Wiley & Sons, New York, USA.

Karselis, T. (1973) *Descriptive Medical Electronics and Instrumentation*, Charles B. Slack Inc., Thurefare, N.J., USA.

Khandpur, R.S. (1990) *Handbook of Biomedical Instrumentation*, Tata McGraw-Hill Publishing Company Ltd., New Delhi, India.

Klein, B. (1974) *Introduction to Medical Electronics*, Foulsham-Tab Limited, Slough Bucks, England.

Kline, J. (1976) *Biological Foundations of Biomedical Engineering*, Little Brown and Company, Boston, USA.

Mckay, R.S. (1968) *Biomedical Telemetry*, John Wiley & Sons, New York, USA.

Mohan Murali, N. (1982) *Biomedical Instrumentation and Hospital Systems* (Vol **1**), Monograph, Osmania University, Hyderabad, India.

Pfluger, T., Reinfelder, H.E., Dorschky, K. and Oppelt, A. (1987) Comparison of various filtering methods in digital image processing, *Electromedica*, **55**(3): 78–85.

Reilly, J.P. (1992) *Electrical Stimulation and Electropathology*, Cambridge University Press, Cambridge, England.

Strong, P. (1973) *Biophysical Measurements*, Tetronix Inc., Beaverton, USA.

Webster, J.G. (Ed.) (1988) *Encyclopedia of Medical Devices and Instrumentation*, John Wiley & Sons, New York, USA.

Weibell, F.J. (1974) Electrical Safety in the Hospital, *Annals of Biomedical Engineering*, **2**: 126–148.

Welkowitz, W. and Deutsch, S. (1976) *Biomedical Instruments, Theory and Design*, Academic Press, New York, USA.

Yanof, H.M. (1972) *Biomedical Electronics*, F.A. Davis Company, Philadelphia, USA.

Index